THE SCIENCE OF
GAME OF THRONES

THE SCIENCE OF
GAME OF THRONES

HELEN KEEN

CANCELLED

CORONET

First published in Great Britain in 2016 by Coronet
An imprint of Hodder & Stoughton
An Hachette UK company

First published in paperback in 2017

1

A CIP catalogue record for this title is available from the British Library

Paperback ISBN: 978 1 473 63234 9
Ebook ISBN: 978 1 473 63233 2

Typeset in Adobe Garamond by Hewer Text UK Ltd, Edinburgh
Printed and bound by Clays Ltd, St Ives plc

Hodder & Stoughton policy is to use papers that are natural, renewable
and recyclable products and made from wood grown in sustainable
forests. The logging and manufacturing processes are expected to
conform to the environmental regulations of the country of origin.

Hodder & Stoughton Ltd
Carmelite House
50 Victoria Embankment
London EC4Y 0DZ

www.hodder.co.uk

ha zhavvorsaan erin ha timviroon eshna meanha
avineserak, ma san athzhilaroon ma athchomar

CONTENTS

CHAPTER 1

HERE BE DRAGONS

Where we discover whether dragons are entirely the stuff of legend; how real-life dragons have sex, whether they could actually fly, and the truth about fire-breathing.

A world with dragons is a world where nowhere is safe. The greatest defensive fortress ever built in all of Westeros was Harrenhal. To the north west of King's Landing, in the Riverlands, Harrenhal sprawled with its thick walls and high towers, its great hall blazed with the warmth of 35 hearths. But before our story even begins the proud towers have been melted like candles, the defensive walls are charred and broken, and many have died and left no trace because of the force that rained down on Harrenhal – dragonfire. A dragon breathes fire with the intense heat and strength of a blast furnace. A juvenile dragon can burn a man alive within seconds. Grown larger, a dragon will easily take on a warship, explosively breathing flames with such power that the ship isn't just set ablaze but blown apart from the force of the shockwave.

This all seems like the stuff of fantasy, but is it? Well let's stop wondering and take a look, beginning by exploring the strange and mysterious origins of dragons and lizards in our world . . .

⤳ Arrested Development ⤳

When George RR Martin and Parris McBride married in 2011 their wedding gifts from the *Game of Thrones* producers included one of the three dragon eggs used in the show. Unfortunately, we have no record of how Mr & Mrs GRRM reacted to the egg ('No, it's lovely, really, but we did specifically ask for a non-stick wok . . .'). They may have barely noticed it while they were delighting in the afterglow of their by-all-accounts very lovely nuptials – at which *no-one died or anything*. But in order to push our narrative forward, let's imagine that a single question drifted gently through their minds, like smoke from the charred remains of someone Dany's Drogon has just taken against: 'Could we ever ride a dragon of our very own?'

In short, could life imitate art?

We begin the Game of Thrones when dragons (and magic) have long been absent from the Seven Kingdoms. According to Westeros' ancient lore, dragons gradually declined and finally died out, with Maester Pycelle helpfully informing us that their skulls line the throne room of the Red Keep in order of birth. The oldest, biggest skull, that of Balerion, could swallow an ox whole, whereas the last skull of the last dragon was barely able to manage a chicken nugget.

The 150-year-old dragon eggs that Daenerys is given to celebrate her marriage are a valuable curiosity. They are pieces of portable wealth that she is expected to sell at some point to further the cause of her House Targaryen, but nothing more. However, as is often the case in the Seven Kingdoms, all is not quite as it seems. After her husband's death, Daenerys orders a huge bonfire to burn several objects and people, including a magical 'wise woman' (who Daenerys holds responsible for her husband's demise), herself and her dragon eggs. Remarkably, from this fiery destruction Dany wanders out, singed but very much alive, and we discover that the eggs have hatched into three adorable baby dragons.

Now, intriguingly, eggs that hatch after a long delay are not entirely unheard of in our world. Reptiles in their eggs can experience what is known as 'arrested embryonic development'. This may suggest a sitcom about an interrelated bunch of socially dysfunctional fetuses, but the term actually refers to an unhatched reptile that essentially presses the 'pause' button on its own development and waits until more favourable environmental conditions arrive (e.g. burny-burny-magic-y-fire, in our particular *Game of Thrones* case). Researchers believe that two main factors have contributed to the evolution of this remarkable process – it's speculation but, both of these factors might particularly apply to the eggs of our fictional dragons. First, it's something that happens particularly to eggs with very thick shells, and second, to offspring who don't receive much in the way of parental care (we can't imagine Dany's Drogon, for example, being super-nurturing). However, typically this developmental 'pause' only lasts up to about a year in reptiles, compared to the 150-year intermission of the spark of life of our dragons, so perhaps not.

Sex and the Single Dragon: A Beginner's Guide

Once you actually *have* some dragons it might be relatively easy to make more? Obvious sex differences don't seem to be a big deal in the dragon world as we know it: males generally grow a bit bigger but it's hard for the human eye to tell dragon-ladies from dragon-gentlemen. Nevertheless, observations of dragon mating have been made by singularly determined biologists: they testify that, in the wild, Komodo dragons – the non-flying, non-fire-breathing kind of dragons we have in our world – will seek out chances for sexual reproduction, given the right conditions, and despite their solitary nature.

Let's travel to the Komodo's native Indonesia, and watch . . . When an opportunity for socialising with the other sex comes up – say, while gathered around the messily slaughtered carcasses of their prey – Komodo courtship can occur. It just isn't going to make for a cute viral YouTube hit anytime soon. The male dragons often kick things off with some bipedal wrestling. Their bout of wrestling can go on for several days and the females are expected to watch and look interested.

All good things must come to an end, however, and eventually a male will win over a female's favour. After licking her scales attentively for some time, he reveals his hemipene. Which – if you wish to imagine it – is a sort of double-headed penis. He produces it/them from a sort of pouch called a 'cloaca', like a magician producing a bunch of flowers from up his sleeve. Ta-da! Or, if you must have the scientific and appropriately serious description, he 'everts his hemipene', and impregnates the, by-now, well-licked and possibly ever-so-slightly bored and wrestled-out female dragon. If all goes according to plan, she lays some fertilised eggs, at which point Monsieur Dragon seems quite eager to forget the whole sorry business ever happened and pays no attention whatsoever to the resulting offspring.

Alone, the female dragon will repel potential predators and guard her nest containing the vulnerable eggs as they develop – a task so stressful biologists have speculated it accounts for why female Komodos don't grow as large as males, and also why they die much younger. Researchers observing the Komodos over a period of eight years found plenty of males leading active lives, fighting and mating into their 60s. The researchers also found, or rather couldn't find, any females older than 33, (sort of like a Hollywood Rom-Com casting, but with lizards).

Once the eggs have hatched, Ms Komodo leaves them to it and gets back to her life – what's left of it. The young dragons must toughen up from day one and fend for themselves, usually by hiding up a tree until they're big enough to avoid being eaten by the other dragons – including their forgetful, hungry Ma and Pa. Ah, Komodo family life!

～ Virgin Births and Flaming Hot Eggs ～

But what about Dany's dragons? Given that they are, as far as we know, the only three dragons in the Known World, can they reproduce and if so how? George RR Martin penning a dragon sex scene? I think we'd all be curious to see *that* . . .

Like their distant cousins, the Komodos, might Dany's dragons be able to reproduce sexually (with each other) or even asexually (alone)? As Chester Zoo keepers discovered in 2006 Komodo dragons are not at all like picky and hard-to-please pandas when it comes to reproduction. The world's largest living lizards are endangered and down to their last few thousand in the wild. As a result a Komodo in the family way is a particularly welcome and happy sight. But Flora – Chester Zoo's Komodo dragon – was unusual in that she had become pregnant without any contact with a male dragon, ever, at any point in her life.

When her eggs hatched as all-male baby dragons, it was discovered that the young lizards' genes were entirely derived from their mother's own biological makeup, although they were not exact clones of Flora's. This is known as virgin birth or, as biologists call it, 'parthenogenesis' (from the Greek *parthen* meaning 'virgin' and *genesis*, meaning 'well-known progressive rock band whose songs, let's be honest, can drag on a bit'). Flora's babies caused no end of excitement in the world of sexy dragon studies and beyond, with *Scientific American* proclaiming that this curious phenomenon may even provide 'one explanation why Jesus was not a clone of Mary'.

From the get-go, dragons (in fact, many reptiles) are not quite like us when it comes to reproduction. In humans and most other mammals, the X and Y chromosomes determine biological sex at conception, or even before. Females have two of the same kind of sex chromosomes (XX), and males have two distinct sex chromosomes (XY). However, as Jennifer Harrison, lizard specialist explains, Komodo dragons work with a ZW chromosomal system and it's the females that have the mix (ZW) and the males that have two of the same (Z) chromosomes.

An unfertilised Komodo egg gets a Z or a W chromosome from the mother. Usually the father will supply a Z, making some children ZZ and some ZW: males and females. Perfect. What the virginal zoo Komodos are doing, Harrison points out, is doubling the chromosome in their unfertilised egg. So the unfertilised Z becomes ZZ (male) and the W becomes WW (fails to develop). Parthenogenesis in Komodo dragons *always* results in male offspring; you never get a mix of ZW asexually. Also, if you're a lonely female looking to occupy a new ecological niche, producing males is the best strategy ever. You can create males to reproduce with and keep the line going. If that's an adaptive evolutionary strategy, it's a brilliant one!

The Maesters of Westeros in their wisdom tell us that dragons are as 'changeable as flame' – gender fluid even – meaning they can magically choose to become male or female at the drop of a (probably burning) hat. As it turns out, reptiles and dragons in our own world don't quite have that power, but some do have rather a neat system for deciding on male or female offspring, and Mother of Dragons Daenerys, with her flaming egg-hatching technique, should probably take note . . .

Even though they are eager to tell you many details about their pregnancy and upcoming childbirth that you haven't necessarily asked for, many human parents willingly remain in the dark about the sex of their unborn child. 'We don't want to know!' they'll declare. 'We want it to be a surprise!' Of course, it's best to respect this and change tack to something more neutral. (For example, say, 'What if it turned out you could give birth to, say, a litter of puppies or something, that would be wonderful, wouldn't it?' and then speculate about the pros and cons of bringing various breeds and sizes of dog into the world, whether they join in or not. Probably don't say, ' Yes! A surprise! Have you seen *Rosemary's Baby*?')

But even if dragons could have those kind of pre-birth sex conversations, they almost certainly wouldn't, because there's an intriguing temperature scale which determines whether, for instance, a young Australian bearded dragon, native to south-eastern Australia, is male or female.

Scientists have found that some Jacky dragons with ZZ chromosomes (genetically male) come out female. Whether the chromosomal signature ZZ is also directly responsible for a super-long beard, astonishing musical longevity and technical prowess with a guitar, is yet to be determined.

What's going on? Well, some species of dragon can bypass chromosomes as sex-determiners altogether and instead respond to temperature while still unhatched. With Jacky dragon eggs, when incubated hot hot hot (well, 30–33°C) they hatch as females, and also when the eggs are chilled to 23–26°C female dragons will appear. In the middle, er, lukewarm temperate band, that's when you hatch male dragons.

But in the curious world of temperature-dependent sex determination, even these rules aren't fixed. When it comes to the eggs of crocodiles and alligators, males will hatch in hot conditions, while females will emerge from cooler eggs.

Why does this happen? Well, like so much in the world of reptiles and dragons, it's still deeply mysterious – no-one's entirely sure. It would be neat if a parent dragon (or even Daenerys Targaryen) could decide the sex of their offspring by incubating the eggs at different temperatures, but we don't have much evidence this happens. Forty years ago in the journal *Nature*, Eric Charnov and James Bull put forward the theory that parents will adapt to local environmental conditions depending on how males or females of their species fare. So if you're a dragon, the chances of having plenty of choice in meeting and reproducing with your scaly soulmate are informed by climate and temperature.

~ Dragon Flight ~

So, we know much more about the sex or not-so-sex lives of dragons and lizards of our world, but what about the more unusual, fantastic elements of dragon anatomy? Are huge wings and flaming mouths really beyond the realms of possibility? Is the quintessential *dragony-ness* of dragons really disbarred by nature? Could huge, fire-breathing lizards with an affinity for silvery-violet-platinum blondes soar through the skies of Essex as well as Essos? Have they ever?

On the face of it, this does seem to be an easy one: of course not. Dragons are magic, and magic isn't real (apart from Dynamo, he is definitely real *and* magic). But evolution is even more creative than George RR Martin, and just as keen on sex and death. It seems that for every possible dragony trait there may be real-world creatures that could happily perch on Daenerys's unsinged shoulder.

For instance, our world can totally do the big flying lizard stuff . . . we're just about 70 million years too late. Fossilised remains of *Quetzalcoatlus northropi* – a huge pterosaur with an estimated 10 metre wingspan – have been found in Texas, North America. Thought to be the largest creatures ever to take to the air, and weighing in at up to a quarter of a tonne, these ancient reptiles must have been a terrifying sight to behold, even on land where they would've stood tall enough to look a giraffe straight in the eye.

If you happen to visit Crystal Palace Park in south London today, you'll notice it contains some 19th century pterosaur-esque statues that look, well, kind of dragon-ish. (And if you can visit you definitely should, because antique reptiles in all their peculiar, – 'really? Victorian's thought they looked like *that*?' – glory are a wonderful thing, even if, er, *Jurassic Park* it ain't.)

Earth's atmosphere may've been slightly denser all those millions of years ago when dinosaurs roamed the Earth, hence it was a bit easier for heavier creatures to fly about. That's one theory. But both our fictional dragons and the factual 70-million-year-old pterosaurs have a similar problem: how exactly did they (or even could they) get off the ground? Could they really fly higher than an eagle? And, if so, what of the wind beneath their wings?

Pterosaurs have a certain amount in common with birds. For instance, their durable hollow bones are strengthened by internal struts. Pterosaur skeletons are tough yet light – one of the most robust things to ever evolve. If this extraordinary beast really flew, it hefted its bulk aloft and through the air with bones whose walls were as thin as playing cards. It seems so unlikely, and yet the latest thinking is that these lizards flew thousands of kilometres, reaching speeds of up to 120 kilometres an hour, then gliding for 90 kilometres an hour, far above the dinosaurs that roamed the Earth below them.

But . . . how? Well, for a start they definitely couldn't just 'take off', like birds do. In fact, some believe they couldn't have taken off at all. Japan's Professor Katsufumi Sato of the University of Tokyo travelled to the nature reserves of the remote Crozet Islands in the southern Indian Ocean, close to Antartica, to study the largest birds alive today. Sato believed the upshot of his study showed that any creature weighing over 40 kilos – twice the weight of our current biggest flying bird, the wandering albatross – simply couldn't keep flapping enough to stay aloft, particularly if it hit any kind of bad weather. Hmm.

But Professor Sato's conclusions about weight, flapping and flight are disputed by those who definitely believe pterosaurs could fly, who believe that pterosaurs could, pretty much, touch the sky. And they've been thinking about it every night and day (with apologies here to R Kelly). As well as theories about there being a denser atmosphere on Earth back then, differences in anatomy and physiology need to be taken into account too. Experts believe pterosaurs probably wouldn't 'flap' much, they'd soar on thermal currents, like eagles and other large birds we see today.

We cannot ignore the schism that runs through the realm of pterosaur experts with regards to how this flight may have happened. The camp is divided between the 'they ran and jumped off a cliff' theorists, and their fierce adversaries, the 'they did this sort of straight-up jumping, thrust-y thing' believers. At the moment, the thrust-y corner are certainly making the most convincing case. And despite universal agreement that real pterosaurs didn't really look *all that much* like dragons, if you watch *Game of Thrones* carefully, next time one of Dany's dragons takes flight, you may well see it doing this very same thrusting skywards move I have knowledgeably dubbed 'the straight jump-up', just like one of their real-world bygone cousins. (Though they are also able to soar off cliffs too. That's the privelege of being magic.)

This isn't accidental. The special effects team behind the TV show have talked extensively about how, even though they're creating a fantasy show, they want the audience to believe in and feel the organic 'reality' of the unreal dragons. Joe Bauer, the Emmy-award winning visual effects supervisor for the show, reveals that he and the crew have extensively studied and modelled their dragons' behaviour on real-world creatures – birds, bats, prehistoric creatures and reptiles – including, of course, our friends the Komodo dragons. For instance, they spent a considerable amount of time working out when the young dragons in the early series would start to develop a 'threat pose' and what exactly that pose would be like. Making the fantasy reality wasn't easy. 'We started with a lot of George RR Martin's ideas about dragons,' series creator DB Weiss revealed, 'because George has definitely spent more time thinking about dragons than anyone I've ever met.'

It's possible though that paleontologist and biomechanist Dr Michael Habib from the University of Southern California could give George RR Martin a run for his money. Dr Habib is more than a little obsessed with the anatomy of giant flying reptiles. He worked out that if pterosaurs were trying to fly by 'gliding' off like a flying squirrel, given how they were built, take-off would involve them dislocating their hips every time they took flight, thus dealing a major blow to the 'they flew from cliffs' camp. Pterosaurs also wouldn't have been great at moving about on the ground. In fact, they would probably have moved like a bat – that is, a bit weirdly. Go look up a video of a bat walking along the floor – odd, right? But again kind of reminiscent of how we see Dany's dragon Drogon move and sometimes take flight on *Game of Thrones*.

How do we know this? Well, we're millions of years too late to observe pterosaurs in their natural state, but we can feed the information that we do know about them – from fossil evidence of size, wingspan, bone density, super-strong legs and so forth – into a computer, and see what's possible . . . Just such a computer model by Dr Habib showed that pterosaurs were only ever a powerful hop away from flying. They could then spread out their wings and, with a strong downstroke, generate the lift needed to take flight. Vampire bats take flight in the same way. Take-off would be a very important consideration for dragon rider Dany. We've seen her dragons run-then-glide to take wing. The more plausible jumpstart for a creature of this size would involve the great hope of House Targaryen really having to hang on to her Drogon for dear life.

~ Fire-breathing ~

Fire-breathing is less easy than flight. It's not really something that occurs in nature, but we do have a much smaller equivalent, though it works, kind of, in reverse. The bombardier beetle sprays a mix of hot, noxious chemicals from its abdomen when it's feeling threatened. It stores two compounds (hydrogen peroxide and hydroquinone) in its body, mixing them together with water and catalytic enzymes in a separate internal chamber to create an unholy hot, gassy, explosive spray.

But if tiny creatures farting bleach doesn't quite capture the dragon-y grandeur of ancient Valyria, what are our other options? One might be the, er, power of actual farts. As some readers will know, many animals produce an abundance of flammable flatus. At the risk of over-egging the potential of colonic gas, if you couple it with a spark, perhaps via a pulse of electricity (hat tip, the electric eel who can generate 600 volts plus) you're definitely getting warmer in your hunt for dragon breath.

I'm truly sorry to go there but human wind contains methane, hydrogen sulphide and hydrogen, and can be relied upon to explode enthusiastically with the addition of two friends, three bottles of hard cider and a cigarette lighter. But even with some sophisticated biological rerouting so that it's coming out of the right end, it's clear that humans, like most other animals with just the one stomach, can't manufacture gas in the quantities necessary to burn down the towers of Harrenhal.

This is an area where the ruminants, with their multiple stomachs full of methane-creating bacteria, have definite advantages. A cow can produce between 250 and 500 litres of highly flammable methane a day and, as it happens, most of this is belched out. In 2013 it was reported that a build-up of methane from a particularly afflicted dairy herd, coupled with an accidental spark of static electricity, 'nearly blew the roof off [the] barn' in Rasdorf, Germany (though doubt has subsequently been cast on the ability of even the windiest cow herd to achieve this through gas alone).

So we're not exactly talking Balerion the Black Dread here, but it may be that one day, who knows? After a lot of genetic tinkering, a Daenerys Targaryen in our world will be able to ride valiantly into battle to claim what's hers on the back of a genetically modified fire-breathing heifer . . .

But where does the idea of fire-breathing come from? It's often supposed that our early ancestors' ideas of dragons come from finding the fossilised remains of dinosaurs, and quite reasonably imagining the terrifying flesh that covered these great bones, in all its fierce glory. Countries that are rich in dinosaur finds, such as China, England and Wales tend to also have a rich mythology of dragons. But that still leaves us with a puzzle. Given that we simply don't find fire-breathing in nature, where does the *idea* of dragon fire actually come from?

A number of intriguing and wildly different theories have been put forward over the years.

The first and most obvious one to rule out: 'It looks a bit like they are breathing fire . . .' Komodo dragons, for instance, flick out their forked pinkish tongues, which . . . looks a bit like a flame? If you squint? Could the story have got started because our forebears saw dragons sticking out their tongues, and embellished? This is certainly possible, but it makes our ancestors seem a bit like easily frightened myopic dummies. So maybe not.

Anthropologist David E Jones puts forward a meatier and more intriguing theory about the origins of dragons in his book *An Instinct For Dragons*. Interestingly, he posits that our terror of these beasts is far older than human history itself. He believes that a fear of certain predators is passed down through evolution. As we modern human sophisticates update our Facebook status and pick up a Starbucks, something from our origins as scurrying, hairy, tree-dwelling creatures in a hostile world of predators still stirs within us. A faint hint of leathery-skin-slither, just heard as we fall asleep; a half-remembered guttural roar as we brunch. It's there, lurking.

Jones notes that African vervet monkeys get distinctly rattled about three particular kinds of predator. They give out a specific warning cry when they spot either serpents, big cats or birds of prey. So, what do you get if you cross a snake and a lion with an eagle? *A dragon*! Dr Jones answers, enthusiastically, in his (very serious) book. The mental image of a winged-roaring-scaly dragon therefore becomes a life-saver of a mnemonic for things that you (and the vervet monkey) really, really want to avoid being grabbed by. His argument is that the fear of dragons has somehow become hardwired into our imagination by evolution, passed down through a million and more generations and hence we still find it easy to conjure up our own frightening 'brain dragons'.

For many millennia, our mammalian ancestors lived in a threatening world where every day they were presented with potentially dangerous situations that required them to make a single basic choice: fight or flight? These days, of course, when we modern humans need to make even the most pressing decisions, there are always more options – such as fight, flight, or kick back for a bit and procrastinate by checking Facebook and those wedding pics from that girl your brother used to work with, while sadly regretting your own life choices . . . But it was *different* back then in our prehistoric days. Those potential ancestors of ours who didn't 'fly' in fear of the unholy trinity of lion-eagle-snake were less likely to live long enough to pass on their genes to their descendants; one of the aforementioned predators would eat them before they had a chance to breed. Hence the cautious and fearful creatures would live to shriek and run away (and have similarly nervy offspring) another day. Survival of the scarediest.

Of course, most of us live very different lives motivated by very different fears today. (At least I hope for your sake you aren't reading this to kill time while keeping an eye out for hungry lions; for goodness sake *concentrate* if you are.) However, even though it's not necessary, our bodies hang on to things just in case, kind of nostalgically, from earlier stages of our evolution. Just look at our coccyx, a tail bone from back when we all had tails. or male nipples, from back when men did all the breast-feeding. (That last one is definitely true.) Of course, there's no way to know for certain if this dragon-fear theory is correct. But I must make a mental note to never show vervet monkeys that Disney animated *Robin Hood*, where all the villains we know and love from Sherwood Forest are played by lions, pythons and vultures. It's sad to know they will never be able to relax and enjoy Friar Tuck heartily portrayed with gusto by a badger.

So how does fire-breathing blast across this warning from our furry tree-dwelling past? As Jones notes, even when fictional dragons don't actually breathe fire, their breath is always described as noxious – poisonous, smoky and hot. While not actually fiery, the experience of being suddenly, unexpectedly close to the open mouth of a big cat predator, its hot breath steaming in the cold morning air, reeking of

24

dangerously fresh meat, would have truly terrified our vegan monkey forebears. But . . . well, this doesn't seem to quite add up. I'm a vegetarian, and I've stood at close quarters having a conversation with someone who's recently wolfed down a Big Mac, and while it's really not great I've never yet reached for a fire extinguisher. While Dr Jones's theory is intriguing and in some ways highly persuasive, it just doesn't fully answer our question about real fire-breathing . . .

So . . . what if it *was* real fire? What if people believed in frightening dragons breathing fire because they witnessed the blast of fiery breath with their own eyes? Matt Kaplan, in his book *Science of Monsters*, explains that this is what could have happened. Dragons are often imagined to live in caves, or under the Earth, and be guarding treasure. When our forefathers ventured underground, to mine for resources or even to raid ancient burial mounds for valuable relics, they would certainly have encountered natural gas – specifically methane, the highly explosive chemical compound that ignites ferociously with the introduction of the flaming candles these underground explorers would've carried to make their way in the dark. Kaplan recounts a story told in the Middle Ages about the 5th century warlord and leader of the Britons, Vortigern.

A castle in Wales kept collapsing, so Vortigern asked his advisors what he could do about it. They recommended finding a boy with no father and sacrificing him so that his blood could appease whatever was causing this construction headache (thus setting the scene for the most grisly episode of *Grand Designs* ever). The 'bastard boy' they found was brought before King Vortigern, but he told a different story of the castle – that there were dragons fighting under it and hence to rebuild it somewhere nearby. The king wisely spared the boy's blood, took his advice and they never looked back; and the boy grew up to be Merlin, the greatest magician in all the land. Though, as Kaplan points out, the area in Wales where this nightmare construction is said to have taken place is rich in coal deposits, so there could well have been issues with the king's palace and natural gas. Maybe Merlin was not just a great magician, but also a canny geologist on the side?

This idea of underground gas explosions giving rise to the likes of Drogon may seem a little far fetched, but when you combine this with the idea of bones of dragon-dinosaurs emerging from underground, perhaps even after a dramatic landslip, it's certainly a nice neat theory.

Capturing the Targaryens: the Blood of the Dragon

The unpredictable behaviour of the Targaryen dynasty has had huge political ramifications in Westeros over the years. Could this have been due to some of their more unconventional family traditions? While nefarious brother and sister act Jaime and Cersei Lannister are surreptitiously doing it all over the place, one family that doesn't bother to hide its incestuous tendencies is House Targaryen. They wear their inbreeding with a pride seldom seen outside Norfolk. (OK, sorry, seldom seen outside a rather obvious comedy sketch about Norfolk . . .)

The Targaryens encourage incestuous marriage to conserve as much of their special 'blood of the dragon' as possible for their offspring. And who can blame them? At least some of the time it gives them special powers: the power of heat resistance, a rapport with flame-breathing reptiles, and super-fetching, shampoo ad, silver-blonde hair. All this has worked out pretty well so far, at least for Dany, but what happens when our royals keep it in the family? How does all this genetics and bloodline stuff work in the world of Gregor Mendel (the father of genetics) as opposed to the world of Gregor Clegane?

The problem that arises from breeding with blood relatives (apart from the inevitable awkwardness at Christmas) is that it can give recessive gene variants a chance to cause a visible effect. We all inherit two sets of genes, one from each parent. We can carry all sorts of potentially harmful recessive gene variants, passed down from either our mother or father, with no ill effects whatsoever, just as long as they're matched with a harmless dominant gene variant from the other parent. Autosomal recessive disorders, like sickle cell anaemia or cystic fibrosis, occur when both mother and father pass on the same recessive gene. And the chances of both parents carrying the same potentially harmful recessive gene variant increase *significantly* when both parents come from the same family.

Probably the best documented case of inbreeding in European royalty occurred in the 17th century. Charles II of Spain – also known as Charles the Bewitched – was the end product of his family's determination to retain possession of all their property by marrying each other. This 'endogamic matrimonial policy' aka keeping it in the family meant over a period of 200 years, nine of the dynasty's eleven marriages were between blood relatives. Charles's mother was the niece of his father, and his grandmother was also his aunt. Not a single outsider married into the Spanish royal line after about 1550 – and when King Charles was born over a hundred years later in 1661 they'd reached an acme of inbreeding.

Charles himself was constantly ill, physically weak and prematurely aged, he had great difficulty speaking. Without an understanding of genetics, doctors at the Hapsburg court were baffled and unable to explain Charles's many health problems. The Spanish Inquisition then stepped in, and an exorcist was called upon to ask Satan himself – was the King bewitched? Of course as you might expect the answer came back 'yes!'. (A witch's spell involving the liquefied brain of a murderer and a cup of chocolate was blamed.) The potions and rituals Charles was prescribed against this unholy possession made his health even worse, however. Charles died, childless, at 38. A King wouldn't usually be subjected to an autopsy, but because it was widely believed Charles had been hexed there was much interest in his death – medicine was given another shot at explaining what had happened. A record of his post mortem examination makes rather grim reading:

'a very small heart of the size of a grain of pepper... the intestines putrefactive and gangrenous... a single testicle as black as coal and his head full of water'.

Charles's death as the 'last Hapsburg King of Spain' provoked European dynastic upheaval, leading to the War of the Spanish Succession (1701 – 1714). Thousands of lives were lost as rival powers fought for control and influence in the vast Hapsburg Empire.

But inbreeding isn't necessarily all doom, gloom and 'this chocolate tastes *weird*.' On the flip side, there's evidence that, over time, inbreeding can actually purge a population of the effects of harmful recessive gene variants. These 'bad' genes are way more likely to show their effects, so, ultimately, the lines of the carriers are more likely to die off. Thus while the results of successive generations inbreeding is generally bad for the particular individual, it's often good for the population as a whole. Following what experts think was a population bottleneck several thousand years ago, cheetahs are all closely related, but also have remarkably few genetic illnesses. However, a lack of genetic variation in a population could also mean they're all susceptible to catching and subsequently dying from the exact same disease. Individuals simply aren't different enough to stand a chance of diversely reacting to disease and surviving.

So, is it possible that Robert Baratheon could have wiped out the Targaryen line most effectively during the War of the Usurper, not by joining forces with Ned Stark and Tywin Lannister but instead by infecting them all with a particularly nasty bout of flu? I put this theory to University of Aberdeen's gene-supremo Dr Jonathan Pettitt, who is politely sceptical. (Not just re the Targaryens, but also to the theory's application in our world.) Dr Pettitt points out that it's difficult to know if the reduced genetic variation increases the risk of extinction, as such populations are also small and having a small population alone increases your risk of going extinct. If there's not many of you, simple bad luck can finish you off as a species, regardless of your genetic diversity. If Daenerys is pretty much the only living, confirmed Targaryen we know about, should we be worried about her untimely demise? Well, remember the scene in Season 6 of *Game of Thrones* where Dany sets fire to the Khals at the temple of the Dosh Khaleen, emerging unburnt and nude and somehow all-at-once channelling Joan of Arc, Carrie and Kim Kardashian? I don't think we – or science – should worry too much for now.

CHAPTER 2

FIRE POWER

Where we find out how Wildfire is all too real; the long-lost secret of Valyrian steel; a tried-and-tested golden formula for killing someone who's annoying you

----

~ Wildfire ~

In the climactic Battle of the Blackwater, the mighty forces of Stannis Baratheon are ranged against the beleaguered capital, King's Landing. Tyrion Lannister, occupying the unenviable position as the Hand of King Joffrey Baratheon (best not to think about where Joffrey's actual hands have been) sends a single ship to meet the invading fleet. That ship is laden with wildfire and, ignited by a single arrow from the bow of Tyrion's friend the sellsword Bronn, it swiftly engulfs the closely packed enemy galleys in flames.

Wildfire is a creatively unpleasant addition to vicious Westerosi wars as it burns everything before it and spreads like, well, wildfire. After the death of the last dragon, wildfire was developed at the behest of the Targaryen dynasty as a substitute for dragonfire. The manufacture of the substance is a secret closely guarded by the Alchemists' Guild. It is apparently known in less enlightened parts of the kingdom as 'pyromancer's piss'. (N.B. If your pee is capable of burning men alive/setting ships ablaze, please put this book down and seek urgent medical attention.)

Human beings being what they are, it's no surprise to discover that, in our world, science has created a weapon that is a match for wildfire's sinister flames. In the 1940s a team led by organic chemist Louis Fieser and based at Harvard University found that adding a thickening agent to fuel created something that burned longer and also tended to stick to surfaces. Napalm (named after two of the constituents of its thickening agent, naphthenic acid and palmitic acid) is a burning gel, usually based on petroleum, that sticks to roofs, furniture and skin. Being oil-based it will burn on, and not be easily extinguished by, water – just like our Westerosi wildfire. Napalm's horrific role in the Vietnam war is notorious, and the UN finally outlawed its use against civilians in 1980.

But during World War II there was a plan to deploy napalm that could've come straight out of the imagination of George RR Martin himself. The US military plotted to arm bats with napalm. Really. Killer napalm bats! This plan was known as Project X-Ray (as, presumably, Project BatShit didn't quite strike the right note). Professor Fieser was asked to create tiny incendiary devices that the bats could carry into enemy territory. He came up with oblong, nitrocellulose cases filled with thickened kerosene with a small time-delay igniter cemented along one side. This curious (to say the least) scheme originated in the mind of a 60-something Pennsylvanian dentist, Lytle S Adams, who wrote to President Roosevelt vividly setting out how 'the millions of bats that have for ages inhabited our belfries, tunnels and caverns were placed there by God to await this hour to play their part in the scheme of free human existence, and to frustrate any attempt of those who dare to desecrate our way of life . . .' The president was quick to get on board with Adams's idea of bats as divine defenders of the free world, emphasising to the military, re the crusading dentist, 'this guy is not a nut'. Mmm!

In our world, as in Westeros, this highly potent fire isn't too fussy about what it burns, and during a test flight the bomber bats accidentally set a US military base ablaze. In the event, the deployment of their fatal flames during World War II was prevented by the speedier development of an even deadlier device, the Atomic Bomb. Perhaps in comparison to that ultimate weapon, the wars and tactics in King's Landing aren't quite so terrifying after all.

GREEN EYED MONSTER – WILDFIRE IS BACK

After sitting out a few Seasons, wildfire made a dramatic return to Westeros in the finale of Season 6. Cersei's response to her trial and all her enemies assembling in the Great Sept of Baelor was certainly ... explosive. The Great Sept was the largest building in King's Landing, a huge and holy stone edifice comparable in size and significance to St. Pauls Cathedral in London, or Hagia Sophia in

Istanbul. Many years ago, caches of wildfire had been left underground all over the city by the Targaryen King Aerys II, Robert Baratheon's predecessor on the Iron Throne. By igniting these deadly reserves under the Sept, Cersei's problems conveniently disappear in a blast of green flame – and what a blast of green flame! Is there anything similar that could cause such a spectacular explosion – green or otherwise – in our world? I spoke to Dr Rory Hadden, Rushbrook Lecturer in Fire Investigation at the University of Edinburgh – how would he analyse the great fire of Baelor's Sept? The challenge to finding a real world match seems to be that this isn't really a 'condensed phase explosive' such as you for instance might see in a military situation. Such explosives tend to not produce flames but rather a huge over pressure (the shock waves over and above normal atmospheric pressure that also happen after, for instance, sonic booms). What we witness in King's landing is a detonation, but it also has elements of a deflagration (the term for most of the 'fire' we see in our world) – lots of 'flameyness'. When we see wildfire destroy the Sept, we're seeing an unusual combination – there's 'a huge overpressure *and* we see there are lots of "flames" which suggests wildfire – perhaps like the dragonfire it was designed to stand in for – has some magic in it,' according to Dr Hadden. Fire scientist and engineer Dr Guillermo Rein agrees – commenting that while wildfire can be classified as a 'reactive phosperant liquid' it 'possesses explosive and burning properties not known to any chemical species I can think of. It seems to be a mixture that contains both the fuel *and* its own oxidiser – hence very explosive – but also burns in what looks like diffusion flames which means the fire also reacts with *something* in the surrounding air.' He adds 'I was going to say oxygen, but maybe it's 'magical' chemical can also make nitrogen react?' As nitrogen is the most abundant gas in our atmosphere that kind of magic might be even more terrifying than the look of satisfaction on Cersei's face as she sips her wine in the green afterglow . . .

⌁ Temper, Temper – The Mysterious ⌁ Science of Getting Your Sword *Just Right*

We have all been there. One minute you are making amazing steel weapons that can slice through armour, hold an edge forever and wreak havoc on the world, the next, you forget how to produce the stuff and that's the end of that.

It was exactly the same in Essos. For over 5,000 years a small group of highly skilled wizards used magic to craft Valyrian steel into swords and daggers. The pinnacle of expertly-crafted weaponry, these fine blades were forged in dragon's fire and developed a fearsome reputation for being unbelievably strong, sharp and able to withstand the most extreme of temperatures. Oily black and covered in thousands of intricate and elaborate swirls, they quickly became the weapon of choice for serious-minded dragon warriors across the Seven Kingdoms. Then along came the Doom of Valyria, the wizards lost their magical touch, and suddenly aforementioned dragon warriors were forced to – I don't know – wrap tinfoil around a sharpened stick? Embarrassing.

I became intrigued by the secrets of real world sword-making – and in particular, the process of hardening and tempering the blade of a sword – after coming across the legend of world saviour Azor Ahai and his flaming sword, Lightbringer. Westeros legend tells that to fight the forces of Winter and the White Walkers, and win the Battle for the Dawn, Azor Ahai was tasked with forging a sword like no other. But after numerous attempts to do this – tempering according to the more regular rules of *Bladesmithing for Dummies* – he only produced swords that broke. In the end, he fixed his blade by plunging it into the heart of the thing he loved most of all, his wife Nissa Nissa. She died (Boo!) but the world survived (Hurrah!). Emotional reactions may have been more complex at the time, particularly Nissa Nissa's, but let's press on. We've a sword to get right.

Now the Seven Kingdoms is in need of another hero. Melisandre proclaims that the 'Prince that was Promised', aka Azor Ahai reborn, is Stannis Baratheon. She gives Stannis a 'magic' sword that flames like the fabled Lightbringer, but she is not able to get the sword to radiate heat. It's all just one of her tricks rather than *real* magic. (This may have given Shireen false hope. Ach! *Melisandre!* Why did you have to burn the screaming, sweet, bookish kid with the skin condition?)

The idea of quenching a sword in human blood sounds fantastical, but George RR Martin was undoubtedly inspired by the gory mythology surrounding the creation of Damascus steel, used during the crusades in Europe across the 11th century. The formula for making Damascus steel remained a closely guarded secret with armourers in the Middle Ages. But according to writings found in Asia Minor, wonderfully superior swords may have been tempered in the blood of slaves. In one ancient account, a worthy blade needed to be heated until it glowed 'like the sun rising in the desert' before being plunged 'into the body of a muscular slave' so that his strength would be transferred to the sword.

Even in our modern era, past myths about the process of making great steel are invoked. 'Arabs Tempered Steel By Plunging Into Slaves, New Method Is Better' a mid-20th century headline trumpeting a newly developed cooling oil declared.

One of the neat things about ancient heat treatment is that they had no timers or thermometers, or understanding of the processes, so a lot of it involved experience, closely held trade secrets and a good share of superstition – which may be where the stories of unfortunate slaves came from. (Let's hope!)

Three Steps to the Perfect Blade

'In order to know about magic steel, we first need to know about real steel,' explains Canadian materials scientist and Valyrian steel afficionado Ryan Consell. Techniques for heat-treating a sword are complex, but the process always involves the following three steps:

1. Soaking. This involves bringing the metal up to a high temperature – over 700°C – and holding it there for some time, often hours. The best temperature is very specific to the alloy.

2. Quenching. This is the part where you cool the metal off quickly to achieve extreme hardness. Usually this is done by dunking the steel into a liquid, like oil or water. The rate at which the metal cools off is very important. It defines how the crystal will be structured and what precipitates might form. Even the difference between ice water and room temperature water can make a difference. Oils are the most common quenching medium for blades. They have just about the right rate of heat dissipation. It is not *impossible* that body temperature flesh and blood would provide an appropriate cooling rate for some alloys. Though Consell adds, 'It seems a bit unlikely and certainly hard to get a good, reliable, even quench *just* by running someone through, but not impossible.'

3. Tempering. Often quenching will over-harden the blade, so it is necessary to raise the temperature of the blade and hold it for a while. Tempering at a relatively low temperature – around 200°C – can ease some of the internal stresses and make the blade a bit softer but much less prone to shattering.

∼ The Secrets of Valyrian Steel ∼

Game of Thrones gets swords. Unlike some fantasy history epics we shan't name, the show generally offers a realistic portrayal of both the weapons themselves and how they're used for fighting. Experts tend to agree on this. For starters, a guy who is a professional bladesmith says so on YouTube (and he has plaited his beard like a badass Illyrian so *I believe him*). I double checked, of course.

In a quest to uncover the long-lost secret of Valyrian steel, Ryan Consell spent years immersing himself in the lore of Essos. He began his investigation by examining the properties of ordinary steel, which is made by mixing a large amount of iron with a small amount of carbon. The good news is that the resulting alloy is extremely hard and, as anyone who has tried to make a jelly dagger will tell you, the harder the material, the easier it is to sharpen. There is, however, one major problem. The harder the material, the more easily it shatters; Consell quickly discovered that an ordinary steel sword would run the risk of shattering on an enemy's shield.

Consell then turned his attention to spring steel, which is made by mixing iron with a tiny amount of carbon, and then adding silicon and manganese. Unlike ordinary steel, spring steel can both hold a sharp edge and be remarkably strong. Consell thought that he was onto a winner until he realised that whereas Valyrian steel performs well at extremely high temperatures, spring steel would simply melt away when faced with even the tiniest of coughs from a fire-breathing dragon.

Undeterred, and ever the scientist he moved onto his third candidate: air-hardened steel. The good news is that air-hardened steel isn't fussy – it doesn't need careful heat treatment, just to cool off in the open air. The bad news is that, although it can be polished to a dark grey, it can't have the oily blackness or elaborate swirls commonly associated with Valyrian steel.

Just when he was about to abort his quest, Consell had a brainwave. Maybe, he thought, Valyrian steel isn't steel at all but rather a 'metal matrix composite'. These amazing modern-day super materials are created by embedding ceramic compounds into the structure of a metal. Whilst analysing the properties of various composites, Consell came across a remarkable material called titanium-silicon carbide (or, as researchers sometimes refer to it, TSC). Any blade created from TSC would be amazingly sharp, strong and capable of performing well in even the highest of temperatures. Not only that, but TSC's remarkable structure means that the blade could be made to look jet black and even have Valyrian steel-esque swirls of grey running through it says Consell. 'Combining the properties of different elements in tightly controlled ways which could make something that would seem borderline magical to a medieval society.'

After years of trial and error, has Ryan Consell finally uncovered the secret of Valyrian steel? Alas, we don't know for sure, but the signs are good. He was last seen heading towards the master-blacksmiths of Qohor, clutching a titanium-silicon carbide blade in one hand and patent application in the other . . .

⌁ Waiting for a Star to Fall ⌁
So I Can Make it Into a Sword
(While Listening to Megadeth)

Not all the swords on *Game of Thrones* are forged from Valyrian steel. One of the most exciting-sounding swords in this world belongs to a character we only see in flashbacks e.g in Season 6, Ser Arthur Dayne, aka the Sword of the Morning. (This is the title associated with his House that only one who has proved himself worthy may bear, and you should definitely not be giggling at this talk of 'morning swords' because we're all serious scientifically minded adults here.)

Ser Arthur is a great and famed knight, a Kingsguard who was killed in a fight by noble Ned who has come to find his sister Lyanna Stark at the Tower of Joy. Ser Arthur was said to be the greatest swordsman in the Seven Kingdoms. But not for him the spell-forged Valyrian steel. No! His ancestral sword is crafted from even rarer stuff – it's made from a fallen star. Or, to be more prosaic, a meteorite. The name of the sword is Dawn, and its home is House Dayne's seat in Dorne, at a place called Starfall. This hints at the origin of the sword; how it came to be in the family. Dawn is described in the books rather magically – unsheathed it glows beautifully moon-pale, alive with light.

So, of course, I have to know: are meteor swords a real thing? Do we have them in our world? Well, if you're a regular reader of fantasy literature you'll suspect the answer is a resounding Yes! They even jab through into real life. When the much-loved, much-missed fantasy writer Terry Pratchett became Sir Terry Pratchett he gave a lot of thought to his sword (because knights must have swords, right?). Ser Terry decided his would include a little 'thunderbolt iron', as the meteor-metal is sometimes known. 'Most of my life I've been producing stuff which is intangible and so it's amazing the achievement you feel when you have made something which is really real,' he said of the sword, which he smelted himself. So what is the origin this well-travelled metal? Some of it is from nearby rocks floating around our solar system, some from regions of space far beyond, the molten core of long-gone planets. But it ends up all over our world.

If we travel back in history, we find that from the dawn of ancient times, the world fancied the idea of making things from meteorites. A 'very heavy' statue of a Buddha-like figure, which was stolen by the Nazis from Tibet, comes apparently from an ataxite meteorite – a combination of space rock that is mainly iron mixed with a smaller quantity of nickel. Now known as the Iron Space Man, it was probably carved a thousand years ago, but a chemical analysis reveals it most likely fell to Earth on what's known as the Chinga meteorite field in Mongolia between 10 and 20 millennia ago.

To the Ancient Egyptians, meteorites were 'sky metal' – a precious gift from the heavens. They lacked the technology to smelt iron – which due to iron's high melting point of 1,538°C requires very hot furnaces to get it out of the ore – but highly skilled metal workers made beautiful objects from the iron they found. When King Tutankhamun's tomb was opened, he was discovered in possession of a dagger made from meteorite iron. He also had some iron beads that had rusted, and become dull and ordinary over time, but would have had a stunning rainbow lustre when new. The king's necklace was crafted from space-iron that had mixed with other minerals at high temperatures in a planetary core millions of years ago, before journeying to Earth.

In the 1600s, Mughal Emperor Jahangir was given a breathtaking gold inlaid dagger made from meteoritic iron – a gift from a tax collector keen to seek his favour. The tax collector described how he had to wait patiently for the hot metal to cool, after watching the rock fall from the skies – presumably keeping himself occupied with a few tax returns.

Attila the Hun had his legendary 'Sword of Mars' that fell from the skies – a present from the gods. Russian Czar Alexander I was given a space sword by an Englishman who forged it from a meteor that fell on the Cape of Good Hope.

The high value of meteorite-metal crosses vast distances of space and culture. Perhaps the saddest space-metal tale is of the three great Arctic meteorites.

Known as 'The Dog, The Tent, and The Woman' they were greatly valued by the Inuit people, who used them to make harpoons. Unless they traded with Europeans, meteorites were their only source of iron. In 1897, the American Arctic explorer Robert Edwin Peary returned from the far north with the huge space rocks on the deck of his ship, alongside several Inuit men and boys, who ended up living (and mostly quickly dying) in a New York museum, where the meteorites remain to this day.

But if you looked out of the window now, saw a fire ball streak through the sky and a good-sized rock neatly fall just outside, how would you go about making a sword from it? And would it be a good idea in the first place?

Well, for a start, meteorites are sort of like the cosmos's Kinder eggs, in that it's hard to tell what they contain until you've broken them open. And even then . . .? It might not be what you were hoping for.

If it's solid and visible and survived its fall from space, it probably contains at least some iron, right? Well, the answer is Yes and No. Meteorite collector and expert Martin Goff tells me that out of the world's 62,140 known, classified meteorites, only 1,132 of them are iron meteorites, or 1.82%. However, iron meteorites have always been disproportionately over-represented in meteorite collections. This is partly due to the fact that they are more easily recognised as meteorites (if you asked most of us what a meteorite looks like, we would describe an iron meteorite), partly because they are more resistant to weathering, so can survive for much longer before being found. And partly because they tend to be larger and heavier than other meteorites. In fact, the combined weight of all the known iron meteorites amounts to roughly 90% of the weight of all types of meteorites added together!

Nevertheless, materials chemist and space enthusiast Dr Suze Kundu of the University of Surrey isn't sure if she'd make a sword out of a meteorite. Dr Kundu is currently working with nanomaterials for clean energy and sustainable solar fuels, and is clearly pretty enthusiastic about swords. 'I think I'd prefer the Valyrian or Damascus steel sword myself,' she explains, with a glint in her eye. 'They'd be much lighter, so you could wield a much larger sword.' Dr Kundu describes herself as a nano-chemist, 'literally and professionally'; she is exactly four feet ten inches tall and would rather like a sword of a similar height to take her above and beyond the far away land of five feet.

Ryan Consell agrees with Dr Kundu that 'Most meteoric iron isn't very good for swords. It has too much nickel and unpredictable inclusions of other elements.' But he adds, 'That doesn't rule out the possibility of a meteorite existing with an excellent composition. It's just not common.' So, the Daynes could have got very lucky with the source material for Dawn. And of course there's nothing to say that in *their* universe, meteorites don't contain all kinds of intriguing and strange minerals and substances that we don't have in ours.

The Public Astronomer at Greenwich's Royal Observatory, Dr Marek Kukula, is interested in the unusual magnetic properties of meteoritic iron crystals, which could prove invaluable in manufacturing stronger, more efficient magnets for green technologies (for example, those used in electric cars and wind turbines). It seems, says Kukula, that the magnetic properties these meteorites hold date back to the metal's days at the spinning core of another planet, just as the motion in our Earth's liquid outer core is believed to be the source of our planet's magnetic field. From Dr Marek Kukula I heard about the artist and master blacksmith Matthew Luck Galpin, who has worked with meteoritic iron in his forge on the artwork 'Anvilled Stars' installed at the Greenwich Observatory. He tells me that 'strange gasses' came out while he was heating, smelting and forge-welding the meteorites to a near molten state, burning and hammering out the trace elements, 'it certainly fired my imagination'. He added 'I definitely had a sense of magic. . . of other worlds. Both inner and outer.'

If you're interested in bladesmithing with meteorites, the basic process you follow isn't so different from usual. The meteorite is heated to obtain the liquid metal, then carbonised to turn it into steel. To achieve 0.5% carbon, you would then need to chisel the material into small pieces that would fit into the palm of your hand, before heating them to glow for six hours. Then there's the usual heating and hammering that we see in all real and fictional depictions of a smithy, shaping the metal into a rod that can be turned to form the core of the weapon.

Once the meteoric iron is formed into a rod, it is flattened and welded, usually with two other iron rods, to make what's known as a pattern-welded sword. The rods are heated and hammered, in order to shape and remove impurities, and then quenched to rapidly cool and further harden the metal. The layering of different metal alloys from the meteorite combines with the earthly iron, making swirling patterns in the steel. If you want to make a decorative, pretty sword, these patterns are then highlighted with acid etching and polished to bring forth the natural beauty within the contours of the metal layers. You can actually watch videos of this happening, to an insistent soundtrack of – what else? – death metal. (Seriously guys, just once with the sword videos it would be nice to have, I dunno, Nina Simone? Taylor Swift?) But there's certainly some kind of link – the *Game of Thrones* team released a pre-Season 6 video featuring cast members attempting to differentiate between the names of the swords on the show: Oathkeeper, Heart Eater, Dark Sister, and 1980s metal bands: Crimson Death, Savage Grace.

We'll have iron and steel for a while yet, but according to some estimates, some of the elements and metals we rely on today could run out in 50 or 60 years' time. Earth's friendly neighbourhood space rocks – passing asteroids, near Earth objects – contain the things we want: phosphorus, antimony, zinc, tin, lead, indium, silver, gold, copper, platinum, cobalt. When we look to the future it seems we may be mining minerals and rare metals from rocks *still in space*. Once the privilege of kings and warriors and leaders, it seems that one day soon, perhaps many of us will own our own special objects made of space metal. Start coming up with good names now!

~ 'Golden Words He Will ~ Pour in Your Ear'

No matter how popular the character or how complex the back story, from the moment Eddard Stark lost his head in Season 1 we knew that no character was ever far away from a graphic, gratuitous, grandiloquent end.

There's a lot of death in *Game of Thrones*, but sometimes the manner of someone's offing really sticks with you. Take, for instance, the departure of Viserys Targaryen, Dany's ruthless older brother. He shares her delicate features and unusual hair colourant, but not her kind and noble nature, so it's not *too* upsetting when he meets a sticky end at the hands of his brother-in-law, the Khal.

While travelling with the Dothraki, the Targaryen brother pushes his luck once too often. Viserys's impatience to wear a golden crown befitting a king is answered in a brutally ironic fashion when Khal Drogo, the man who calls Dany 'moon of my life', rains a river of molten gold over his whiny silvery-violety-blond head. This is not the sort of thing even 'the blood of the dragon' can easily come back from. Viserys dies, and quite unpleasantly. Worst royal hat since the one Princess Beatrice wore to Prince William and Kate Middleton's wedding.

Now gold melts at 1,064°C, 'it is known'. The temperature of the average campfire – whether it is warming the Dothrakis' deadly bling soup or Bear Grylls – is pretty unlikely to get that high. However, had the thrifty Dothraki slipped some lead in with their gold to make an alloy of the two metals ('pure gold? He isn't worth it!'), it would have melted at a conveniently lower temperature.

Plenty of people have been murdered *for* gold in our world, but have any been murdered *by* it? The delightful idea of delivering a gilded death to a 'high born' or 'grabby' person is also recorded in our history. In 1599 the Jivaro Indians, natives of Ecuador and Peru, saw off a Spanish governor whose greed for taxing their gold seemed insatiable, unstoppable, right up to the point when they poured molten gold down his throat. (The Jivaro remained, subsequently, unconquered.)

The Spanish Inquisition, whose supreme enthusiasm for devising weird and wonderful deaths for people is only now being challenged 400 years later by Mr George RR Martin, was also apparently fond of killing its victims by tipping hot metals down their throats. In Classical times the hapless Emperor Valerian I was the only ruler of the Romans ever to be captured during a military campaign (the Battle of Edessa, in fact, battle fans). His bad luck was compounded by the fact that he happened to be caught by notorious Persian badass Sharpur I. What followed is subject to some debate amongst historians. One theory is that, after many years of getting Valerian to bend down and pretend to be a foot stool so Sharpur could use him to climb up onto his horse, the Persian emperor executed the Roman by forcing him to neck molten gold. Others suggest that this is simply propaganda to make the Persians sound, well, really bad – in fact Sharpur may have had Valerian flayed alive (House Bolton style) then stuffed his remains with straw and propped what was left of him up against a wall for all to see in a nearby temple. What a humanitarian! Don't worry, the Romans also did this kind of thing to their enemies, before you feel too bad for the guy.

In 2003, presumably one of 'those conversations' in a Dutch pathology lab led to an exploration into this style of death. The *Journal of Clinical Pathology* published a paper under their 'Historical Perspectives' banner with the evocative (even-by-the-standards-of-pathology) title 'Molten Gold was Poured Down His Throat Until His Bowels Burst'. Scientists from the Department of Pathology at the VU University Medical Centre, Amsterdam, made a study with liquid metal and a cow's larynx. Thankfully they are at pains to point out that the cow was dead at the time – the larynx was obtained from a local slaughter house. The pathologists pondered the effect that shock, pain and the inevitability of suffocation would have, before concluding, 'We have shown that in the execution method of pouring hot liquefied metals into the throat of a victim, death is probably mediated by the development of steam and consequent thermal injury to the airways.'

Hey, Science, thanks for clearing that up!

∽ Kill it With (Dragon) Fire. Or, ∽ How to Burn an Astaphorian Slave Master

In Dany's quest to build an army and retake her birthright of the Iron Throne, she makes a curious bargain in the city of Astapor. She agrees to give away one of her dragons in exchange for a slave army of the Unsullied. The deal is brokered with a slave trader who fully expects Dany's dragon to become his newest possession. But he has clearly not spent enough time wondering how dragons take to a life of forced servitude. The answer is 'not very well', and within seconds the guy is unequivocally ablaze.

But how would that work? How much flame would a dragon need to flame in order to flame someone to death-by-flame?

If we want a dragon that doesn't just breathe fire but can actually kill with it, some calculations are required. The mastermind behind the flames here is Dr Rory Hadden, Lecturer in Fire Investigation, School of Engineering, University of Edinburgh. So, let's say our friend, the soon to be ex-slaver, is an average adult male of lean build, weighing around 70kg/150 pounds. Adult humans like him are around 70% water, so if he's going to be burned to oblivion, that's approximately 49kg/108 pounds of water we (or rather our dragon) needs to dispense with.

Dr Hadden calculates that if we assume the Astaphorian's body is already at a temp of around 60°C (it's a hot day) then to vaporise 49kg of water requires the following calculation:

$$49 \text{ kg} \times 2257\text{kJ/ kg} = 110593 \text{ kJ} = 111\text{MJ}$$

We see Dany's dragon exhale for around five seconds to achieve this – a pretty impressive feat.

So 111mJ of dragonfire is required – the equivalent of 22 megawatts of power (that's akin to the peak power output of the reactor in a Los Angeles class nuclear submarine!)

The one caveat, says Dr Hadden, is that this is if 100% of the energy from the dragon breath can be transferred to the unfortunate person. In reality, the fraction that is absorbed is much less than the radiation emitted. This is because the heat transfer from the flames (gas) to the person (solid) is not very efficient. So to be safe, a dragon might have had to develop the ability to produce four or five times the amount of energy.

Say it with me one more time – a dragon is not a slave!

CHAPTER 3

WEAPONS OF WESTEROS – A (BATTLE) FIELD GUIDE

We get licensed to kill: the art of poisoning; living and dying by the sword; the truth about shadowy assassins; sexy armour, and what it means to be a real warrior woman.

~ Swallows & Antidotes, or, Another ~ Man's Poison, or Quick! The Antidote!

Ah the *Game of Thrones* deaths by poison! A source of misery for some and joy for others. GRRM has given us some wonderfully juicy descriptions of the preparations and effects of toxins in the Known World, and we have poison to thank for that strong contender for everyone's favourite death scene – Joffrey getting his come-uppance. The particular poison that does for Joffrey is called The Strangler. Oberyn Martell, long-skilled in the secret ways of 'It must've been something he ate . . .?' describes the process of making this lethal potion from a rare plant that grows across the Jade Sea, noting that it makes the victim's face turn purple. Well, yes, we all enjoyed watching that.

But in the world of *Game of Thrones*, magic can be used to fight the most lethal of poisons. This is not the first time that we have encountered The Strangler – remember when King Stannis's elderly Maester Cressen attempts to do away with the Red Sorceress Melisandre by slipping the poison into her goblet of wine and proposing a toast to the Lord of Light? Prepared to give his own life to protect his Lord Stannis from the Red Woman's influence, Maester Cressen takes a big gulp of the poisoned wine and Melisandre delicately, but decisively, follows his lead. The Strangler quickly seizes hold of Maester, and within moments he lies bleeding from the nose and mouth. For all his learning, Maester

Cressen has no way of withstanding his own concoction. But the last thing he sees as the poison violently overcomes him is Melisandre's serene, untroubled face. The strange blood red ruby choker at her throat throbs with light as it purges the poison out of her body. (Her choker can do even more than that, as we know from seeing Mel without it on.)

The idea of precious jewels protecting you against deadly toxins was wildly popular during the Middle Ages. While death-by-poisoning might be a mercifully rare occurrence today, it was a common fear in medieval times. Certainly, we know how seriously kings and queens took the threat of poisoning. Elizabeth I, now celebrated as ruling over England during a great flowering of culture and commerce, wasn't universally popular in her day. She could easily have appeared in the Tudor equivalent of Buzzfeed's '10 Monarchs Everyone's Trying to Poison', with numerous bizarre plots to kill her being recorded. In one Oberyn Martell-worthy episode, several members of her court arranged for a lethal poison to be smeared on her horse's saddle. However, one of the poisoners confessed all, prior to the queen's daily canter, and the conspirators ended up spending time in the Tower. Both Elizabeth and horse emerged unscathed.

The inventiveness of such poisoning plots was matched only by the bizarre beliefs about how to detect and deal with lethal potions. Edward VI had a ruby and sapphire encrusted toothpick that he believed would 'sweat' if touched by poison. Staying with the Tudors, many believed that diamonds were so powerful they didn't even need to be in the mouth to 'warneth of venime' and that just popping them under 'the left arme pit' would do the trick. Similarly, it was said a sapphire had 'so contrary a nature to poisons' that if you put one in a glass with a spider, the spider would die.

Bezoars – from the Persian word for 'antidote' – are calcified deposits that can develop in the stomachs of both animals and humans (if you do happen to get one yourself, the recommended treatment is to drink bucket-loads of Coca-Cola to dissolve them). They were once seen as highly prized cures for all poisons. Elizabeth I kept one on a bracelet, with a contemporary description of her jewels noting that 'The most parte of this stone is spent', suggesting that it was used on a regular basis. If you couldn't get your Middle Age hands on a bezoar then the alternative was a unicorn horn. These precious objects – actually the horns of narwhals brought back by sailors from the edge of the Arctic oceans – were seen as the best way to detect poisons. (The same is true in Westeros, where the horns of narwhal-like creatures from the Lands of Always Winter are passed off as unicorn horns, as no-one dares to travel to the island of the Skagosi where *actual* unicorns are to be found.) In our world, stories were told of unicorns dipping their horns into ponds and causing all of the muck and unpleasantness to magically rise to the surface, leaving the water clean and pure. Much prized, once again Elizabeth I had at least two pieces of horn, whilst Mary Queen of Scots also had one.

Of course, all of this sounds fantastic, but did they actually work? Well, when it comes to unicorn horns, er, no. But with bezoars, there is good reason to be less sceptical. One of the forefathers of modern surgery, Ambroise Paré was curious and decided to conduct an experiment. Paré was the barber surgeon to several French kings, which gave him easy access to a palace or two. In 1567 a royal chef was sentenced to hang for stealing silverware, but Paré instead persuaded the man to be poisoned and then given a bezoar 'for science'. Unfortunately, the experiment was a complete failure and the chef died a painful death. But more modern-day investigations suggest that both the chef and Paré were unlucky. When a bezoar is immersed in a solution of arsenic the toxic compounds arsenite and arsenate (worst twin names ever) can react with it and become harmless. Although if your drink *doesn't* contain arsenic, then you've just chucked a lump of partially digested gunk and hair from an animal's stomach into your glass for nothing. You win some, you lose some.

⤳ Death by Sword ⤳

Ice is the ancestral greatsword of House Stark. Its dark, smoky Valyrian steel blade is wider than a man's handspan, and it stands taller than Ned's true-born son, the teenaged Robb Stark. Ice's spell-forged steel has held its edge for 400 years, a replacement for an even older sword of the same name. But ye olde sworde is no whimsical antique gathering dust above the fireplace. Because Valyrian steel is relatively light, it is possible that the Starks have wielded the mighty Ice in battle at some point, though we get a proper look at it first in Ned's hands as he uses it to dispense justice to a deserter from the Night's Watch (a sworn brother who has taken the black but now stands accused of separating himself from his vows, so we watch Ned separating his head from his body).

It's traditionally a highborn person's right to insist on trial by combat (with or without a self-appointed champion) if they're accused of a crime. The gods then decide on your guilt or innocence, which means it doesn't particularly matter how good a fighter you or your champion are. Nevertheless, sword practice is a big part of life, just to be on the safe side.

But despite all the practice, the Starks don't have a great deal of luck with sword fights, frankly. Ned inherits the sword Ice after the deaths of his father Rickard and older brother Brandon at the hands of then Targaryen ruler, Mad King Aerys. Coming into conflict with Aerys, who accuses him of treachery, patriarch Lord Rickard Stark duly requests trial by combat – he's a great swordsman. But this only works so long as everyone sticks to the rules. Aerys decides – somewhat unsportingly – that fire will be his champion, and proceeds to roast alive the Stark patriarch in his armour over a blazing pyre as the royal court looks on.

Meanwhile, Rickard's son Brandon has been brought into the throne room with a noose around his neck and a sword that could cut his father down placed just out of his reach – Brandon strangles himself to death trying to reach it. Oh Mad King Aerys! (The clue is in the name . . .) In *A Dance with Dragons*, we learn that excellent swordsman Brandon Stark has told his lover Barbrey Ryswell that he hones his sword so that it is sharp enough to remove a woman's pubic hair. (N.B. Do not try this at home.) Suffice to say Brandon's untimely death at the hands of the Mad King means that his line of feminine depilatory products never really got off the ground.

Ned Stark is also denied any kind of trial by combat, and ultimately endures a surprise public beheading with Ice, his own ancestral sword. Those who live by the sword die by the sword too. But not always as they would expect.

When Bad Executions Happen to Good People

When we watch executions by sword in *Game of Thrones* they tend to go really rather smoothly. Luckily we don't usually have to watch what happens when executions go bad. Firstly, these slick executions are often due to the keen and widespread use of legendarily sharp Valyrian steel blades. Secondly . . . it's because they're not real.

Historically, blade-y decapitation seems to have been a much more fearful and gruesome business. When the English king Henry VIII had his second wife Anne Boleyn executed, he turned to France, England's oldest frenemy, to find an expert swordsman. The executioner was brought over specially to do the beheading with one clean fast stroke, as an act of 'mercy'. Beheadings with swords (or, worse still, axes) were apparently very prone to going unpleasantly awry, and it would regularly take several attempts to part head from body to the satisfaction of all involved.

In 1789, when the ruling French establishment was in the midst of a particularly chaotic and deathly spell (even by their standards), a doctor and medical reformer named Joseph-Ignace Guillotin realised that Something Must Be Done. Possibly this was after watching an execution akin to the time Theon Greyjoy bungled the beheading Winterfell's Ser Rodrik Cassel, the very master-at-arms who'd taught him how to use a sword. (What a time to find out your pupil really wasn't paying attention.)

Dr Guillotin was actually opposed to the death penalty but hoped that bringing in a humane mechanical instrument of execution for all (at that time 'commoners' were often hanged, while the wealthy and influential were decapitated) would be a first step towards doing away with state-sponsored killing in France altogether. It's fair to say that things didn't quite turn out as Dr Guillotin planned. For a start, even though he didn't invent it, the guillotine continued to bear his name, even after his family sued to try to get it changed. The guillotine has the dubious honour of being one of humanity's most enduringly popular *machines de mort* – the last public execution in France was in the 1930s, and was witnessed by a young British actor called Christopher Lee. The final execution of all by guillotine carried out (privately) by the French state was in 1977, just after the first *Star Wars* film came out. There are many contemporary reports of recently decapitated heads reacting with a range of emotions – from horror to indignation – as they were held aloft by the executioner. Could these be true? The current medical consensus is that, yes, possibly you could remain conscious after losing your head, perhaps for as long as 13 seconds. A clean decapitation wouldn't instantly kill your brain activity, but the subsequent lack of oxygen, glucose and other chemicals would. So, just for a moment, it may even be possible for the head to see the neck and body it's just been separated from…

~ Death by Sword-Fighting Wounds ~

Even if you win a sword fight, it can still cost you your life. An injury that became infected was a common cause of death in societies without antibiotics or a strong understanding of how to keep a wound clean. General life advice would include, don't fight Oberyn Martell, the Dornish Red Viper, who earned his nickname after a duel to 'first blood' as a teenager. After inflicting a minor wound, he was declared the winner but, unknown to his opponent, he had hedged his bets by using a sword daubed with poison, so his opponent's wounds festered and eventually killed him. But even a mere flesh wound that literally is a mere flesh wound can be the death of you. Just ask Khal Drogo, who sickens greatly after sustaining a slight wound after otherwise effortlessly winning a fight, or The Hound, left to (almost!) die on a hillside after a minor wound that didn't really concern him putrefies.

Up until the late Middle Ages, it was widely believed that diseases of many different stamps were caused by bad, smelly 'contaminated' air or *miasma* that circulated, particularly at night. But in 1546, the theory of germs getting into a wound and causing infection and possibly death was first mooted by Veronese poet and physician Girolamo Fracastoro, who proposed that it was tiny organisms, or 'spores', that carried the contagion. (A rather grand portrait of Fracastoro painted by Titian hangs in the National Gallery in London today. The painting was said to be given as a payment for treating the artist for syphilis.)

By the end of the 19th century in our world, germ theory was rapidly gaining acceptance. So much so that it gave a curious twist to a duel fought between two high-born aristocrats. In 1892 in Verduz, Liechtenstein, Princess Pauline Metternich and the Countess Kielmansegg had a disagreement over flower arranging. The two had never exactly been bosom buddies, but after this latest spat, they took the surprising, though not entirely unheard of, decision to arrange a duel. This was a proper, highly regulated affair, with a pair of seconds (also female). And the medical professional on hand was, unusually for that time, also a woman – Baroness Lubinska.

Lubinska liked to keep abreast of new medical developments and was a relatively early adopter of the strictures recommended by an acceptance germ theory. If you're injured in a duel by an opponent's sword on a clothed part of your body, their sword could easily catch a bit of your clothing and jab it right into the wound, which could result in sepsis, or 'blood poisoning' setting in. The Baroness's solution was proposed and agreed – in order to minimize the dangers of infection, the Princess and the Countess would both duel *topless*. And all at once a fight over the best way of displaying your lupins goes from being not very *Game of Thrones*, to very *Game of Thrones* indeed.

I'm Too Sexy for My Armour – Beyond the Physics of Boob-plate

Despite copious nudity, one way in which *Game of Thrones* refreshingly differs from some other fantasy shows is when women are armed and fighting in battle they wear 'real' armour to protect their bodies, and not the sort of armour that seems to exist just to display as much flesh and breast as possible. In extreme cases this sort of feminine armour is just a metal bikini that wouldn't do much to protect you from the midday sun, let alone a sword-wielding opponent. This kind of armour is sometimes referred to as a 'boob' rather than 'breast' plate. Of course if you want to wear revealing armour and it makes you feel good – why not?! Sexy armour that unmistakably highlights the wearer's assets has a long, hot history. We have many examples from different periods . . . though they are all for men. The Ancient Greeks of our world had full-on, ripped six-pack bronze body armour, which presented their torsos in quite an eye-opening, and frankly buff-tastic way. Even we prudish English had a go, looking at designs from the designs from the Tudor period, we see that plenty of it was all about unmistakably sexual display. Lusty old Henry VIII even had armour made featuring a *massive* codpiece.

But this armour, certainly king Henry's, seems to have been mainly worn for showing off like heavy metal peacocks, rather than for battle. The primary consideration for armour is what you're going to be doing in it and what – if anything – other people are going to be hitting you with. Armour aficionado Ryan Consell who we met earlier on Valyrian steel comments that we rarely see images of historical women in armour as they were restricted from fighting.

He formulates this relationship re boob plate which neatly expresses some of the problems involved in wearing it

$$Defense \propto \left(\frac{Exposed\ Skin \times Cup\ Size}{Comfort}\right)^{chance\ of\ falling\ out}$$

The biggest and perhaps counterproductive issue if you fancy fighting in booby armour is as soon as an opponent strikes your armour, the force of the blow is channelled into the divot – the

'cleavage' indent between the breasts – which vibrates hard into your sternum and will probably end up shattering it, which . . . isn't pleasant.

As for the armour of our *Game of Thrones* heroines, Consell is generally impressed. For instance the Ironborn's would-be leader Yara Greyjoy has pretty good, realistic armour – though Consell notes she is missing the one piece of armour that was one of the first to be worn and the last to be discarded – neck protection. Brienne of Tarth does better on this front – she's clearly well armoured all over, in fact Brienne's armour is pretty much perfect for a woman who is a knight, and she appears both powerful and attractive.

Still, I can't help thinking the *Game of Thrones* armourers may have missed a trick with not arming the boys with massive codpieces . . .

Warrior Women

Yara 'Asha' Greyjoy, Brienne of Tarth and Arya Stark are three of the fiercest fighting women in all of fantasy literature. Warrior women thrive in the Seven Kingdoms. But in the real world, the role of females in combat situations is more controversial. 'Should women serve in war?' is seen as a hot topic for television debates, news bulletins and talk shows. And the generally accepted view is that it wasn't until the 20th century, when women had achieved political agency, and started working outside the home, that it became acceptable for females to fight in the armed services.

As it turns out, warrior women have been around for longer than you'd think. And there are plenty of parallels between the women of *Game of Thrones* and famous female fighters in history.

Yara Greyjoy, Theon's sister, is a high-ranking member of the Iron Born, whose culture bears certain striking similarities to that of the Vikings. In the absence of her brothers, she is something of an oddity in her society, allowed to live a more typically 'masculine' existence by her father, the Greyjoy Iron Born leader. So much so that she briefly hopes to become his successor and the leader of the Iron Born herself.

If we picture a Viking fighter, we probably picture a man, probably in one of those annoying horned helmets – which they never wore IRL (apparently). This would almost certainly irk Freydís Eiríksdóttir, anti-heroine of the Viking sagas, and daughter of that other notable Viking ass-kicker Erik the Red. According to the Sagas, Freydís made it her mission to become a morally compromised fighter who you would not want to cross. In one story, for example, we witness the moment she, and some male Viking comrades, capture some enemy women. The men don't know what to do with them. 'Kill them!' instructs Freydis. And when no-one listens, she rolls her eyes, fetches an axe and proceeds to demonstrate that a woman can commit what we'd now refer to as heinous war crimes every bit as well as a man can.

Later, pregnant, she sails to the New World in command of her own men, aboard her own ship. Her Viking party encounter some native Americans and retreat in fear. Freydis is appalled at their cowardice and – eight months gone by now, so possibly slightly less handy with an axe – makes a terrible war cry, brandishes her sword about, and then, finally, pops one of her breasts out. The native Americans are duly terrified at this last gesture, and flee.

(One has to ask – why did her breast have this effect? Was it a bit like on *Friends* when Monica flashes Joey? Was it merely the audacity of the gesture that bought time for the embattled Vikings, or was there something quite scary about her breast – could it be mistaken for an angry shark? Or had she drawn something weird on it? (No, of course not. No-one ever does that when they're a bit bored. Certainly not.))

Someone I've Always Meant to Get Around to Reading, who is often described as the Father of History, the Greek writer Herodotus speaks of a race of Amazons: free wheelin', fightin' and hollerin' women who fought and shot arrows from horseback. Imagine the Dothraki Khals, but with Khaleesi's instead. Or Lyanna Stark, who is described as such a fine rider that she must be 'half horse' herself. (Though dead before the timeline of the show and books begins, Lyanna haunts those she left behind, and her secrets may yet turn to influence who finally wins the game of thrones.) Until relatively recently our classical scholars chuckled indulgently at these stories. And Herodotus also set the precedent for Classical male writers discussing female fighters and being weirdly obsessed by their chests. Seriously – I have been reading up on this stuff and Guys! You need to give the constant 'but what happens with their boobs while they're fighting, eh?' talk a rest – it's not as interesting as you think. Really.

Read historical authors on women fighting! They're prone to sidling up to you on any pretext:

'Reading about some women?'

'Yes – really enjoying this bit.'

'Well, if it's a women bit, I expect you'll be wanting to know what their breasts are up to?'

'Not really. I have my own; you get used to them after a while.'

'Let me tell you about her bosom, which—'

'Enough with the boobs!'

Anyway, back to Herodotus, who goes on about the Amazons cutting off one of their breasts in order to fire a bow and arrow better. So far, so probably fictional. But then (boob-business aside) in the later 20th century more and more intriguing archaeological evidence began turning up. Excavations on the Steppes revealed the graves of women buried with spears and knives, bows and arrows. Careful examination of the women's skeletons revealed wear and tear consistent with having used these weapons in life. We also know what they wore: practical trousers (just like Brienne), and pointy hats. They bear a lot of similarities to the women Herodotus describes.

There wasn't anything greatly different about the innate strength of these women compared to, say, modern women. It's just that horseback is a great leveller. If you're on a fast horse with a bow and arrow, the usual sex differences matter less. (This also seems to be the experience when Lyanna Stark takes on the men; it's widely suspected that she disguises herself as the Knight of the Laughing Tree and unseats all three of the other knights who had earlier bullied her friend Howland Reed.)

When it comes to squaring up with men, the average woman is less strong in general terms than the average man. But this doesn't mean that women don't want to or can't fight. Rosi Sexton is a former British Mixed Martial Arts champion who also studied maths (or math, for all you Americans) at Cambridge and Manchester universities, where she gained her PhD. She told me she believes that while 'self defence' is seen as acceptable for girls and women to take an interest in, if you're female and interested in actual fighting as a sport, that's still seen as highly unusual. Defending yourself? Well, that's fair enough, you had no choice. But actually wanting to get into scraps, even in a professional context, is still not entirely acceptable, respectable behaviour for a woman. Dr Sexton also tells me that in her experience training 'evens out' a lot of differences between men and women. Approximately matched in terms of size, a well-trained woman could defeat a less well-prepared man. Evidence suggests that the historical Amazons were real fighters, but their acceptance as regular members of their society may sadly still seem a little fantastical to our modern day eyes, and to Brienne and Arya . . .

Strongest of Them All

———•———

Gregor 'The Mountain' Clegane, the strongest man in all Westeros, can crush a human head with his bare hands. In real life, the actor who plays The Mountain, Hafþór Júlíus Björnsson, has won the title Iceland's Strongest Man for the last five years, and he's also finished in the Top 10 Strongest Men in the World. Pretty impressive. But some physical anthropologists believe that the kind of strength that passes for magnificent today may have been simply 'meh' in comparison with the strength of ancient man ... and woman.

So in the past, was everyone a Mountain? *Was no head safe?* And what about warrior women like Brienne of Tarth, who seems to be able to slug it out with men in battle. Can we find any clues in history to identify women like her?

Measuring strength from fossil bones it is clear that the bones of Neanderthal people grew larger in response to muscular stress. As Australian anthropologist Peter McAllister explains, 'The human body is very plastic and it responds to stress. We have lost 40% of the shafts of our long bones because we have much less of a muscular load placed upon them these days.' In other words, we are simply not exposed to the same loads or challenges that people were in the ancient past, so our bodies haven't developed those dense, strong bones. Even with strenuous exercise, the training of elite athletes today doesn't come close to replicating a day in the life of a Neanderthal.

McAllister believes there's strong evidence to suggest a Neanderthal lady would have 10% more muscle bulk than a modern European gentleman. He explains that she could reach 90% of Arnold Schwarzenegger's bulk at his peak in the 1970s if she trained to capacity (and wow, what a training montage that would be!). It gets better though. She wouldn't just be a challenger. Because of the quirk of her physiology, 'with a much shorter lower arm, she would slam him to the table without a problem,' McAllister says.

Wow. What is best in life? Well, not so much crushing your enemies and seeing them driven before you as hearing the exaltation of their women as they beat you in an arm wrestling contest, it seems! This may not be entirely true. Modern human women also have shorter forearms, but that doesn't seem to give them extra strength over modern human men. Victory in arm wrestling seems to come down to sheer muscle. But still, Neanderthal women would indeed seem to have comparable muscle mass to modern human men – though I guess we'll never know how that arm wrestling would've turned out.

The Bows and Arrows of Outrageous Fortune

When Tyrion kills his controlling, abusive, politically skilful father Lord Tywin, he not only deals a deadly blow to the golden power of his family, the Lannisters, he also radically changes the political landscape of the Seven Kingdoms for ever. The method he uses to affect this great change is a bow and arrow, one of the oldest weapons known to human beings – with the power to change the course of fortunes, battles and ultimately history.

Tywin, of course, was shot by an arrow while sitting on the toilet. He didn't have a chance to prepare for the assault and don battle dress, and so was executed easily. We have records of many people being killed by arrows while out of battle dress, or dying from a shot to an exposed place, like the face. But there is little evidence to suggest that armoured knights – the tanks of the Middle Ages – were killed by arrows penetrating their armour.

Many, many reconstructions of arrows-versus-armour are available to watch online. They are usually narrated, in tones of wistful enthusiasm, by precisely spoken Englishmen who re-live battles in which their countrymen bloodily defeat the French; or sometimes you'll find an affable American gentleman with a very large neck and fruit for targets who makes blunt comparisons with assault rifles and tries not to look too disappointed as he shoots his arrows watermelon-ward.

What everyone seems to agree on, however, is that it is extremely difficult to actually damage armour with an arrow unless it's at close range – less than 20 metres, say. Obviously archers would not often be able to get this close to charging knights on horseback. So, in our Middle Ages, in the regular course of things, you would've been much more likely to get shot with a few arrows but survive (at least for a while) in the manner of Jon Snow after Ygritte fires off a few in his direction, rather than shot at close range like the unlucky Tywin.

A Single Arrow is All It Takes

A bow is a simple device. You pull back the string, energy is stored in it and then released with the arrow, directly at a target. A well-placed arrow is a game changer. Perhaps the most famous WTF? arrow of all time was released by a Norman bowman and lodged in the eye of the hapless King Harold at the Battle of Hastings – thus making the French king, William the Conqueror, ruler of our lands, changing the cultural face of England and ushering into the English language an époque of Frenchification and jokes about bidets.

Harold wasn't the only king to become a target. Richard the Lionheart, aka Richard I of England, was mooching about while laying siege to a castle in France when an archer on the battlements caught his eye. Richard noticed the guy was using a frying pan as a shield. (Some sources say it was a saucepan. Others speak of a George Foreman grill. Detailed culinary classifications do get lost in history's swirling vortex, but it was something kitchen-y and a long time ago. Ok?) Anyway, this amused Richard no end. 'It's so good I'd put my name on it,' he is said to have chortled. And while this was going on *another* French archer (some said just a boy) spotted the king's unprotected state and loosed an arrow, hitting Richard right in the left shoulder. It all escalated rather quickly after that. Back in his tent, dying Richard summoned the crack shot boy/man and forgave him, but the King's retinue were less merciful and once Richard had breathed his last they had the French culprit flayed alive, House Bolton style.

∽ Longbow V Crossbow ∽

As every British person who has ever taken a vague interest in medieval battles knows, the English longbow, also called the Welsh longbow, is famed as a feared weapon which turned the tide of some of history's most popular slaughter-fests. An oft-cited example is the Battle of Crécy in 1346, where the English were hopelessly outnumbered, and also apparently outclassed. Stats: French, knights, lots; English, peasants, a few. Yet the English won. How? Was it really down to the longbow; was that really the most bad-ass weapon of its day? Or is there a certain amount of patriotic misty-eyed-ness going on here?

The history is still hotly disputed. Longbows made of yew certainly had an advantage over crossbows in that the wood itself has curious properties. Yew is hard on the outside while on the inside the wood is supple and bends to a highly *un-yew-sual* degree. Longbows can also be reloaded much quicker than crossbows, and so archers were raining down many more arrows per minute. But these arrows weren't likely to be killing anyone, at least until the soldiers were almost face-to-face. So it seems that the main effect of all those English arrows was to cause feelings of confusion and panic, and generate chaos, as horses weren't particularly armoured and so would be scared and injured by volleys of them, sometimes shedding their knights in the mêlée as a result.

If we allow the longbow its mastery, we must ask, why did the French and the Scots continue to use crossbows and not learn from the English and Welsh longbow mastery? Usually other countries rush in to get a load of something effectively deadly, so isn't that itself a sign that maybe longbows *weren't all that great*? A recent paper has asked this very question and posits that politics is the answer. While England and Wales were generally politically stable in medieval times, France and Scotland were not. Two economists believed that this was crucial, as the English had less to fear from an armed population rising up against the ruling classes, whereas the Scottish and the French wanted to make sure arms were not readily available to their grumbling peasantry.

England was unusual in that royal decrees told every Englishman between 15 and 60 to practise with his longbow on a weekly basis. Football was even banned for a while to keep the population focused. Evidence that at least some Englishmen took this edict to heart comes from skeletons found on Henry VIII's sunken flagship, the *Mary Rose*. Archaeologists have discovered that the skeletons of the archers on board show signs of Repetitive Strain Injury from a lifetime of regularly using heavy bows requiring a 'pull' of up to 200lbs/90kg – the weight of a modern-day washing machine.

Two Fingers Up

It is often said that the charming English habit of 'sticking two fingers up' can be laid at archery's quivering door. Some speculate that this friendly gesture originated as a sign of defiance – Welsh and English archers would wave their two fingers aloft at their French opponents to show they were still very much capable of pulling back the strings of their bows. It was said that the French would mutilate captured English archers out of spite.

Where do we begin? Well, English archers were usually 'low born' – that is, they would not have any value to ransom – so rather than chopping off their fingers the French, had they been feeling vindictive, would simply have killed any captured bowmen quickly with their swords, rather like Ramsay Bolton in the aftermath of a battle. The only original contemporary source referencing the 'finger chop' is actually French, and it refers to the English king telling his fighters on the eve of battle that 'Hey, you know what the French are like, they would chop off THREE of the fingers on your hand if they could'. The French author is appalled by this and makes sure to include a footnote about how outrageously unfairly anti-French this remark is, that's what the English are like, they simply cannot mention the French and not be rude and stereotypical about the French! And then the account ends abruptly because he's gone on strike.

CHAPTER 4

NORTHERN EXPOSURE

Where we break the ice with our friends in the North; Hodor's 'expressive aphasia'; warging and the science of astral projection; the psychology of revenge: Arya Stark's kill list.

~ Brain of Thrones ~

Here's a curious thing: Hodor isn't Hodor. His name is actually Wyllis. Well, in the *A Song of Fire and Ice* books he's called Walder (possibly in homage to his family's liege Lord Walder Frey) but he is Wyllis in the TV show (possibly in homage to the 1980s comedy favourite

Diff'rent Strokes – 'What you talkin' about Wyllis?') Confused? It's not surprising. The World of Ice and Fire echoes to the sound of an enigmatic character repeating the word 'Hodor' over and over and over again. As a result, almost everyone – be they book reader or television viewer – thinks that his name *is* Hodor. But they are *wrong*. It's Wyllis. So, what's going on? Why is Hodor – sorry, Wyllis – so fixated on this single word?

The answer may lie in some groundbreaking scientific research conducted over 150 years ago. In the 1860s, French surgeon Paul Broca was busy studying people's brains, laying the foundations for modern-day neuroscience. During the course of his brainly experiments he came across a middle-aged man called Pierre with the strangest of speech impediments. Whenever Pierre tried to say anything, the only word that he could utter was 'tan'. Unsurprisingly, everyone eventually came to refer to Pierre as 'Tan' in the same way that we all think of Wyllis as Hodor. This terrible condition affected many areas of Pierre's life. At one point, for instance, he asked a decorator to paint his house and ended up with a series of monotonously mid-brown rooms. Just kidding.

Anyway, when Pierre obligingly died, Broca cut open his skull, removed his brain and discovered damage on the left-hand side of his frontal lobes (or, as modern-day scientists snappily refer to the area, the 'inferior frontal gyrus of the left cerebral hemisphere'). Over a century of subsequent work has proved a direct link between this area of the brain and speech production. Whenever the left frontal lobe area is damaged, people struggle to speak – which is why having a stroke on the right-hand side of the brain is generally much better news. And this area of the brain is now referred to as 'Broca's area' in honour of the great Frenchman.

For a long time we were wondering what happened to Wyllis to cause his 'hodoring' behaviour. Perhaps he suffered a stroke or a tumour, or even a blow to the head? (Intriguingly, he does have a little scar on his head, but it's on the right-hand side.) Extensive damage to Broca's area is also sometimes caused by malnutrition, but that seems unlikely in Wyllis's case, given his enormous girth. One thing is clear: Wyllis is clearly exhibiting a severe type of 'expressive aphasia'. He can understand what other people are saying and respond, but he struggles to produce more than a single word.

Broca's work paved the way for our current understanding of the brain. Before Broca had full access to Tan's, sorry Pierre's brain, most scientists were convinced that the brain worked as one big meaty lump to produce people's thoughts, feelings and behaviour. Broca was convinced that this was not the case, and that each part of the brain was involved in a specific task. Discovering that damage to a tiny part of the left hemisphere severely disrupted speech production proved that he was right. In fact, this astonishingly simple idea underlies all of modern-day neuroscience, including all of those pretty brain scans showing the areas associated with vision, memory . . . and orgasms, and provided a crucial insight into what's happening between your ears right now. Hodor? HODOR! Hodor . . . *Hodor*.

⟋ Sneaky F*cker Strategy: The ⟍ Surprising Secret of Samwell Tarly's Success

Ah Samwell Tarly, aka 'Lady Piggy', aka 'Lord of Ham', aka 'Prince Porkchop'; self-confessed coward, son and dispossessed heir of Lord Randyll Tarly of Horn Hill. Many of us *Game of Thrones* fans will have felt for this bookish, clumsy, chubby and socially inept character. His father is a powerful Westerosi lord – one of the Seven Kingdom's noblest and most feared military minds, whose House words are 'First in Battle'. But Sam (as he's affectionately known) blew his right to great wealth and privilege by being hopeless at all things fighting-related. Hence he's forced by his terrifying dad to join the Night's Watch – the alternative being an even more sudden death in a hunting 'accident'. And we all know how those go down (just ask the late King Robert of House Baratheon).

When we first meet Sam, he's the frightened nerd being horribly bullied by the jocks. His new 'brothers' in the Night's Watch are taunting him, calling him 'Ser Piggy'. But it's nothing new. He's been shunned and shamed his whole life for lacking the traditionally macho skills that are prized in high-born young men of his ilk. His father has repeatedly tried and failed to turn Sam into the sort of alpha male heir he was expecting. Lord Tarly's bid for Father of the Year previously involved dressing his son in women's clothes, making him sleep in chainmail, and even enlisting the help of warlocks, who bathed Sam in aurochs' blood. When all that failed, and his second son was born, Randyll Tarly seized his chance to offload Sam, the previous one to the ends of the Earth (the Wall) and forget he ever existed.

Despite this frankly terrible parenting Sam has remained a kind and thoughtful person, one who loves reading books and probably long walks along the beach. Amid the sea of troubled masculinity that is the Night's Watch surging beyond the Wall, he's thrown together with Gilly, a kind and resourceful wildling girl trying to save her baby boy. It's apparent that she's drawn to him and he's drawn to her too – and well, we are cheering them on.

Of course, this is *Game of Thrones*, not your average rom-com. Gilly isn't a quirky waitress with a heart of gold, or the slightly uptight owner of a downtown art gallery with a heart of stone, no Gilly is a traumatised wildling survivor of the hideous Craster's daughter-marrying, son-sacrificing dynasty with I should imagine a rather brave, anxious lion-type of heart. Nevertheless, as Sam observes, the interesting thing about Gilly is that in spite of her abuse by grisly Craster, she remains optimistic about the future.

Sam's sworn oath to the Night's Watch *should* mean he is neither a fighter nor a lover. After solemnly taking his vows, including quite a big promise of celibacy, given his staggeringly obvious lack of skill with a sword, he's assigned (relegated) to the administrative Steward's division of the Watch. Yet, somehow, he manages not only to fall in love but also to have sex with Gilly, *and* to magnificently slay the White Walker who threatens his love and her baby son. Even before Gilly can read, she's super-impressed by Sam's knowledge and love of books – and what he can glean from marks on a page. She tells him he's like a wizard (which he takes as a huge compliment, obviously).

So how does 'unmanly' Ser Piggy manage to stand out and get the girl amid the testy testosterone of the Night's Watch? How does he succeed in an environment that is loaded against success for a young man like him? And can Sam and Gilly's relationship progress?

Science may help us on this one. In particular, the findings of one of the 20th century's most eminent evolutionary biologists, Professor John Maynard Smith, best-known for his theories on natural selection, and his research applying game theory (ideas originally developed to analyse poker and chess games) to the 'game' of being a living creature and getting what you want.

Before Smith came along, evolutionary biologists generally assumed that it was inevitable that natural selection prefers its males to fight their way to the top; that way, the dominant male gets the most chances to pass on his genes to the next generation. For surely the dominant males demonstrate their dominant-ness by doing super-masculine things, like hitting each other and sitting with their legs really far apart on public transport.

Smith realised that there may be more than one way for the fittest to survive and thrive and win the Game of Life. Like humans, animals will watch and try to interpret or predict the actions of others before they decide on a course of action. Thus Smith brought mathematics and probability into formulating and evaluating questions of behaviour in nature. And he argued that, rather than favouring one particular set of behaviours (aggressive dominance), natural selection tends towards maintaining a balance of differing characteristics within a species, so as to maximise the survival of at least some of said species when facing, for instance, a variety of circumstances or challenges in a difficult world. He dubbed this the 'evolutionary stable strategy'.

The part of his thesis that feels custom-made for Samwell is the bit Smith called 'Sneaky Fucker Strategy', which explains why sometimes subordinate males get the girl. Put the term into a search engine, and Google helpfully offers 'searches related to sneaky fucker strategy': 'sneaky male syndrome' (OK, let's call it that from now on), 'klepto-gamy' (fancy way of saying the same thing) and 'actor Clive Owen' (er . . . what?). And surprisingly, at the time of writing, there isn't even a specific entry for the strategy on Wikipedia – kingdom of the esoteri-cally knowledgeable – which, possibly, indicates a conspiracy of silence from some of those who might well be expected to most benefit from said strategy . . .

The behaviour Smith describes is universally recognisable, however. In nature, the titular 'sneaky male' is observed waiting for his oppor-tunity while the alpha males are off fighting, or showing off their prowess. Finding the coast clear, he chooses the moment to make his move on a female. And if she likes the look of him, they can mate while the non-sneaky fighting males are still earnestly scrapping over her.

What's in it for the females? Well, research conducted with 'sneaky male' dung beetles show they often have 'better' genes to pass on to any joint offspring than the fighting males. They can also often be really off-the-cuff witty about the dung beetle equivalent of *America's Next Top Model*.

In human groups, the 'sneaky male' may be the one who genuinely prefers the company and interests of the women around him to those of the men, or who just seems shyer and more sensitive and therefore attracts more potential partners to them. This highlights the radical theory that, for some women, pleasant attention is more enjoyable than being fought over. Imagine?!

So-called dominant alpha males may even be onto their rivals' sneaky strategies. Writing about red deer stags contesting for hinds to add to their harems, Richard Dawkins and John Krebs describe the relative rarity of stags going against each other with antlers on full beam, as it's easy for one or both to become injured, and also because 'subordinate sneaky fucker males' may slink off with hinds if a fight is prolonged.

Prolonged fights are still something of a house speciality in *Game of Thrones*, plus roaring at each other until your opponent backs down. But the unlikely blossoming of Sam and Gilly's relationship gives us a tiny bit of hope in a world where there's very little love or romance. (Unless of course you count Ser Jaime Lannister musing on the things he does in the name of his incestuous love for his sister Cersei, before casually defenestrating the innocent seven-year-old Bran Stark?) But we all hope to go through life finding and feeling love. It's a basic human emotion. Even if it doesn't work out, even if it wasn't real, it's gotta be worth something, *right*? Gilly and Sam are both warm-hearted and kind, but have both been abused and battered on their journeys through life, up until the point they find each other. And I hope you'll agree, finding someone who loves you for you, and makes it so all the other things in life make sense, that's the dream. (That's what gets me out of bed in the late afternoon, anyway.)

So, will Gilly and Sam make it against all the odds? Well, this is *Game of Thrones*, so don't buy your wedding hat just yet. Sam may have easily disregarded some of his vows, but it's hard to see how he can stay together with his lover. He tells Gilly he would prefer to marry her rather than any princess. But no dice, he's in the Night's Watch for life and disappointingly they're all about 'brothers before mothers'. Ho hum.

But equally, never underestimate the 'sneaky' ones. The pair may well be alive and prospering long after some of the more violent characters have competed themselves out of existence.

And it's worth pointing out that at the end of *A Dance with Dragons*, we discover Sam may currently have The Horn of Winter, a fabled instrument that may magically protect or destroy the Wall: that icy barrier which right now is all that stands between the White Walker Others and the realms of men. Sam, who loves old things, even broken old things, is given the curio by Jon Snow after it turns up buried with a stash of dragonglass. Of course, it could just be an obsolete old warhorn, but if it *is* The Horn, well, combined with his Valyrian steel sword, Sam has potentially a pretty big role in the great battle of ice and fire to come, even without the greatest fighting fortitude.

⌒ Skinchanging – ⌒
Out of Body Experiences

Escaping from the flesh-and-blood prisons of our own bodies and effortlessly flying across a blue sky or running swiftly through a cool forest is something many of us have dreamed of, particularly during a bad hangover. But is an out-of-body experience possible? Can we ever be aware of our surroundings when not 'in' our own bodies? How do we truly know where we end and the rest of the world begins? And can we manipulate this knowledge?

In the world of *Game of Thrones*, the secret power of 'skinchanging' (or 'warging') is possessed by a chosen few. This is the ability to slip the bonds of your own body and enter and control another creature, seeing, smelling and feeling the world as it does. Those who practise this skill are called wargs – though strictly speaking (at least in the world of the books), 'warging' is the ability to change into a dog or a wolf, the beginner's level in the difficult art of skinchanging, beyond which few progress.

Though a warg can skinchange with any living creature, it is important not to go rushing in but to take care choosing your animal. The wildling skinchanger Haggon sets out some general advice. Birds are to be avoided, due to the dangerously intoxicating sensation of flying – you'll never want it to stop and will become a bird for ever and ever. (Then before you know it, your nest's mortgaged to the hilt, you're trying to get the chicks into a good local school – jeez, where'd the fun times go?) Forging a bond with a dog, or even a wolf, leads to a trusting warg-relationship that becomes more assured with time. Cats on the other hand really aren't having any of it, so are best avoided, as you might imagine.

There are also rules – a sort of warging code of conduct – that Haggon lays out. It is taboo to eat human meat while in the body of another animal; you must avoid the awkward scenario of being inside another animal when it's mating, and therefore possibly also inside another animal, as it all gets a bit Russian (sex) doll. And hey – two's company, threes's a crowd ... But the worst warging crime of all is to try to control the mind of another human being, not least perhaps as it's frighteningly reminiscent of the control exerted over the army of dead wights by the White Walkers (see page 154).

When Hodor is 'warged' by Bran, we witness this as a horrible trans-gression; Hodor tries to hide inside himself to get away. Bran is an extremely gifted warg but, morally, this is highly dubious.

George RR Martin has said that all the Stark children have the ability to warg 'in their genes', as it were. Bran, one of the younger Stark boys, begins to develop his ability after he falls from the tower and is unable to move around in his own body. He realises that the gift he has is both significant and problematic. Not many wargs have been born since magic fled the World of Ice and Fire – maybe less than one in a thousand children possesses the ability. But still, among the smallfolk of the northern kingdom, any child suspected of being a warg will be left out to die.

The freefolk, north of the Wall, value the skills of the wargs, but even wildlings keep their distance. For, as Old Nan used to say, as she whispered scary stories to Bran in the nursery at night, how do you know if the man is wearing the beast or the beast is wearing the man?

The idea of travelling into the bodies of other animals or humans may seem to have zero parallels in our world but, in fact, every year hundreds of people report 'out of body experiences' (OBEs), or astral trips, that typically involve the sensation of floating above one's body, and often observing one's body from a place outside it. While they don't tend to involve 'entering' other people or animals, in some OBEs people even report leaving their body behind to visit other places and even other 'dimensions' before returning to themselves. Generally, the experience seems to be a very positive one, and some of those who have regular OBEs even learn to control their trip, and actively aim to get into the right state for it. Like warging, it seems to impart a sense of freedom and possibility to those who experience it.

The phenomenon has been much studied and some interesting results reported, but despite claims from some that they are able to leave their bodies at will, apparently no-one has managed to do so in a scientific setting in order to 'prove' their abilities by recalling or reporting specific information. For instance, subjects have not been able to identify objects previously placed on a high shelf that's invisible from the ground, or outside the window on an otherwise inaccessible ledge. And as a result, so far, psychologists can't agree that anything greater than chance has been demonstrated in the lab.

When it comes to 'body swapping', closer to warging of the show/books, two researchers, Valeria I. Petkova and H. Henrik Ehrsson, made the stuff of fantasy literature (and Hollywood comedies) into a distinct area of scientific study with the help of some virtual reality headsets, mirrors and a mannequin.

In their 2008 paper entitled 'If I Were You: Perceptual Illusion of Body Swapping', Petkova and Ehrsson acknowledged the question of just how tenuous our hold on our own bodies and the space they occupy can be, by exploring the imaginative conundrum of suddenly finding yourself inside the body of another. (Yeah – again – not like that.)

Why do you think you're inside your own body? Because you can see it? But what if you looked down and saw another body, utterly different from what you're used to? If you could also touch that strange new body – and feel it being touched by others – then what? From neurological studies we've learned that damage affecting frontal, parietal and temporal lobes of the brain can make you feel like you're outside of your own body, and there are illnesses that cause sufferers to fail to recognise their limbs as their own.

In order to better understand the processes that underlie these experiences, the researchers gave participants VR headsets that relayed all the visual information they received. And to trick the brain into accepting the 'body swap' from every angle, the participants were also initially touched on the torso while watching the mannequin with whom they were 'body swapping' being touched in an identical way.

Petkova and Ehrsson succeeded in creating the illusion of swapping bodies across genders – the mannequin 'appeared male', yet female subjects could still feel themselves inside 'his' body. They also achieved a measurable physiological response when the mannequin was threatened in some way. They even tricked participants into feeling as though they'd changed places with one of the researchers, and shaking hands with themselves. Ehrsson says this illusion is highly convincing: 'The first time I experienced the body-swapping illusion with [fellow researcher Valeria Petkova] I almost started screaming because it was such a surreal and striking experience to shake hands with myself using a different body.'

Focusing on this apparently playful reversal, the paper sets out a response to 'the fundamental question of why we have an ongoing experience of being located inside our bodies' – something which has preoccupied psychologists and philosophers for hundreds of years.

While warging offers an ultimate escape from inside the walls of our own flesh and blood, experiments that show we can fool ourselves into believing inanimate objects are part of us, and how easily we can be tricked to not recognise our own bodies, remind us that our brains are constructing our sense of ourselves 'from the inside'.

In a world that so often wants to categorise us according to how we look from the outside, experiments with the self-defined limits of our bodies can give us a new fluid vision of ourselves, free from other's limiting views of our physicality and our body's capabilities. Maybe in the future, if such technology can be used for entertainment, we'll find ourselves with the same dilemma as the wargs: exhilarated by new experiences, but warned not to stay too long in case we never want to return to ourselves.

EXPERIMENT WITH A FAKE HAND: TOUCHING ON THE SCIENCE OF HOW YOUR MIND CAN LEAVE YOUR BODY

FOR this experiment you will need:

A table

A fake, rubber hand (or an inflated rubber glove, ideally of a similar colour to your skin tone, but this is easier if you come from Springfield)

A large book

A bath towel

2 medium-sized paintbrushes (optional)

A hammer

A friend

Have you ever wanted to feel like your hand is no longer your own, and instead have the feeling of an entirely different hand that gives you a partial out-of-body experience? Well you're in luck! Prepare to warg. (Just a tiny bit . . .)

Oh, and if you don't have a fake hand but you'd like to acquire one, I'm sure you won't regret it. They are available fairly cheaply online, where they are listed, slightly spookily, as 'soft practice hands' (for people studying nail art, last time I looked). Or, the other day, I walked past a shop in London's trendy Soho that had a variety of fake hands complete with forearms in a fetching window display. But then it turned out they were for . . . something else. Anyway!

To begin . . .

Sit down at the table and get comfortable. Place both your hands on the table, relaxed and with palms down. Move your right hand about 20cm to the left and have your friend place the fake hand where your right hand was previously (if the dummy hand is the left hand, move your left hand 20cm to the right, instead). Place a large book standing up vertically between your right hand and the fake right hand, and then place the towel over your right shoulder, using it to close the gap between the fake hand and your torso, in the manner of a sleeve (a towel-y sleeve, but still). Look down and see – does the dummy hand sort of look like your hand? Is your real hand hidden? Good!

Your friend should sit opposite you and begin to gently stroke your real hand and the fake hand at the same time. They can either use their fingers or the two identical paintbrushes if fingers feels a bit too intimate (though if you are interested in someone and they feel the same this could also be a great way for you to finally break the ice together. Think of the story you could tell your grandchildren! And then terrify them by chasing them round the room with your matchmaker fake hand, which you will have nostalgically hung on to for all these years. Awww!)

But back to our experiment. Your friend should stroke both hands in *exactly* the same way – for instance, simultaneously working through each of your fingers and the fake hand's fingers in turn. Once you feel properly relaxed, look down and focus on your fake hand.

After a few minutes something quite strange will start to happen. You will begin to feel the sensation of being stroked, not on your real hand, which is off to the left behind the book, but on the fake hand in front of you. It will probably feel a bit numb and odd, but nevertheless, it will feel like the fake hand is the one being stroked. If you had a way of accurately measuring your skin temperature at this point, you'd notice that the temperature of your real hidden hand was beginning to fall ever so slightly compared to the rest of your body, which suggests that your brain is diverting your blood supply away from this hand as you've begun to adopt the rubber hand as your own. Indeed, if you close your eyes and point with your real, not-being-used-in-the-experiment hand to the space where you feel your hand is, it would point to the fake hand.

Once you've experienced the illusion of the rubber hand gradually becoming like your own flesh, your friend, using their skill and judgement as to timings, should take the hammer and firmly, but not too violently, use it to hit the fake hand. The *fake* hand. JUST HIT THE DUMMY HAND. I'd like to emphasise that bit again, here. (It's particularly important if you wish your friendship to continue beyond the experiment.)

Even *watching* the hammer approach your fake hand should be enough to make you react and flinch, and the hammer strike will probably make you jump. And yet, it's not your hand; it's not part of you! The illusion should promptly cease at this point, as your physical reaction to the hammer blow causes your whole body to tense slightly and brings your attention back to your real, hidden hand.

It's certainly a powerful illusion, and a little disturbing as it proves how easy is it to fool our brains by providing false sensory feedback. This illusion only takes a few minutes to disrupt your body image and your sense of being inside yourself. Your brain 'sees' the fake hand being stroked, 'feels' the stroking and swiftly constructs a sense of 'you' based on this information.

This may be evidence for neuroplasticity – the idea that the brain can change, quite radically, when prompted to by a new experience.

As well as being fun, the fake hand illusion is related to pioneering work carried out by Professor of Neuroscience Vilayanur S. Ramachandran. Ramachandran wanted to study the well-known but little understood problem of phantom limb pain. When a person loses a limb, a foot or a hand (as Kingslayer Ser Jaime Lannister does) a lot of time and patience is required to continue living a normal life. But often this process of adjustment is made all the more difficult by the person continuing to feel terrible pain in the 'phantom' lost limb, even though it is physically gone.

Ramachandran believed that this might be at least partly due to the brain's disorientation as it sent signals for movement to the missing body part, but there was no visual feedback to confirm the movement was happening. As the fake hand experiment demonstrates, there is a very powerful connection between what we see and what we feel.

Ramachandran decided to devise an experiment for people who had undergone an arm amputation. He and his team designed a large mirrored box which you could step inside of and see both your surviving arm and its reflection in the mirror, creating the illusion that the missing arm was back. The participants were then asked to perform a number of actions: for instance, clench 'both' fists or wave 'both' hands, and watch the mirror arm move and react with their real arm.

Now the missing limb was 'moving', responding to instructions *and* the brain could 'see' it. For the majority of those taking part in the study this was enough to reduce the amount of phantom pain they felt – so much so that they wanted their own mirror boxes – suggesting illusion can even diminish the pain of reality.

～ Arya's Revenge – A Dish ～ Best Served Cold by . . . No-one?

Many of us have a list that runs through our heads as we try to fall asleep at night. My list is stuff like 'moisturise shins' and 'stop buying candles shaped like animals'; yours will be different (almost certainly), but we're all familiar with the idea of unfinished business at the end of the day. For Arya Stark, however, that list, the 'prayer' that insistently floats through her mind before sleep, is a recitation of the names of those who have killed and abused her family and friends. We watch Arya grow from a nine-year-old tearaway who rejects playing at being a lady in favour of fighting and shooting arrows, to an accomplished killer, fuelled by grief, anger and hatred.

Diagnosing fictional characters can be tricky, but it seems safe to say that after watching her father's execution and losing so many of her loved ones, Arya is suffering from post traumatic stress disorder. Like some real-life sufferers, she turns to revenge fantasies as a coping strategy to deal with her feelings of helplessness and rage. She decides to train as a professional assassin. And she's still barely into her teenage years. Written down in black and white this is pretty devastating, yet somehow Arya is one of the most lovable characters in *Game of Thrones*.

She desperately wants to find the place she belongs, and the people she belongs with. Ever the wolf-child, she thinks of them as her 'pack'. She remembers her father's words – he tells her not to be too independent, because when winter comes the lone wolf dies, but the pack survives. And we really, really want Arya to find her pack and to be safe. ('Please not Arya. Please not the dire wolves!' goes my mantra at the beginning of any episode of the TV show where they might be in danger of dying or related horrors; like a sort of reversal of Arya's own prayer.)

We watch Arya go through various transformations to safely hide her original identity as a 'highborn young lady'. She loses herself among the 'smallfolk', begging and serving, while all the while becoming a better and better fighter. She loses her sight for a while as bent on revenge, she trains to become an assassin with the Faceless Men – and we remember Gandhi's 'an eye for an eye makes the whole world blind'.

To serve the Many Faced God with the Faceless Men, Arya must put away her House and her name and become 'no-one'; she must have no identity, and yet it seems that all she has left of her original identity is her 'kill' list, and her little sword, Needle – both of which she fiercely clings to.

112

Not long before Arya arrives at the House of Black and White to train with the Faceless Men for a career as a dispassionate killer (which, to be fair, in the brutal circumstances of the world of *Game of Thrones*, does seem to be what any Westerosi careers counsellor might advise for her), she has the chance to cross one of the names off her list.

The Hound, Sandor Clegane, has long served Joffrey and the Lannisters and he kills Arya's friend Mycah without a second thought, when they tell him to do so. Onto Arya's list he goes. But then The Hound captures a runaway Arya, thinks of returning her to her family for some ransom, that er doesn't work out, so the two of them are stuck travelling the Seven Kingdoms together. Arya says she hates him, promises to kill him and yet . . . she doesn't. An unlikely respect develops between the two of them – affection between two birds of a feather, almost. So we see Arya choosing not to act on one of her revenge fantasies. Out of what? Compassion? Fellow feeling? Loneliness?

Revenge fantasy

Brain-imaging scans can perhaps help scientists measure why we get so much pleasure from exacting revenge. Using a technology called positron emission tomography (or PET), Ernst Fehr, professor of Experimental Economic Research in Austria, studied activation in a region of the brain called the dorsal striatum – that's the region involved in registering enjoyment and satisfaction. Fehr studied the emotional dynamics involved for participants in a game of exchanging money back and forth. When one participant made a selfish decision, another player could choose to punish him. And the majority of players elected to do so. The degree of activation of their dorsal striatum region corresponded with how far they were prepared to go. In some cases, volunteers decided to impose a penalty on a selfish wrongdoer even when it cost them some of their own money; those player-participants showed the greatest corresponding degree of activation in dorsal striatum, hence it seems the greatest satisfaction in righting a wrong.

But what about when you can't pay someone back? Are revenge fantasies a healthy means of coping? Do they help us to regain some sense of control over past traumas, or could they lead to further negative consequences? The psychological impact of fantasising about revenge is less widely studied. One experiment incited participants to develop a thirst for revenge 'after they scored low on a difficult anagram task'. Lustily oh oaf pry! (You'll pay for this!)

♫ DERRR
DUUURGH
DER DER DUHHH
DER DER DUHH ♫

More in keeping with our *Game of Thrones*-esque investigations is the study that looked at the occurrence and possible usefulness of fantasies of dishing out revenge as part of a therapy called 'imagery rescripting'. This technique – also called 'guided imagery' – is being used by psychologists to help patients suffering from depression or post traumatic stress disorder revisit or re-create difficult episodes from their past with the aim of updating them. The patient is asked to vividly recall or create a scenario with personal resonance from their past – for example, a time when they were a child and an adult belittled them. They can then enter the scene with the adult and answer back for their younger selves. And redress the balance. For some people this imaginary intervention will be a verbal reckoning. For others it will involve violence towards (in this case) the bad adult.

Psychologists pondered the implications of this. Was imaginary revenge a 'safe' place to go, or could it lead to worse things ahead?

A study published in the *Journal of Behavior Therapy and Experimental Psychiatry* looked at the psychological impact of trauma on mentally healthy participants with no previous history of violent behaviour. The paper's authors created the effects of trauma in the participants by showing them five-minute clips from violent and disturbing Hollywood films where various protagonists were subjected to physical, emotional or sexual abuse. After they'd watched the clips, the participants were split into three groups to take part in different therapeutic exercises.

The first group had to imagine entering the scene they'd just watched and exacting a violent revenge on the perpetrator. The second group intervened in a non-violent way when they entered the scene. And the final group got to magically send the victim to a nice, safe location, like a beautiful beach.

What would make you feel better?

The next day, after they'd completed the exercises, all three groups still felt anger and sadness when they looked at images of the violent perpetrators from the film clips they'd watched. But those who had been violent in their fantasy didn't show higher levels of violent feelings than those who had merely verbally dished out revenge. Though it was noted that the violent revenge sparked more joy for participants than the other two options, the happiest of all were those who had been able to take the victim away to a safe place without saying a word or lifting a fist.

The research results were interesting, if not conclusive (they were dealing with fantasy revenge of fantasy wrongs after all). They seem to indicate that revenge fantasies can be beneficial, or at least not necessarily damaging, so psychologists don't necessarily need to fear someone starting down an Arya-ish road. So, until we invent the 'save them and send them to a beach device', let's not feel too bad fantasising about revenge, while also hoping that the day soon comes when Arya and those like her find their packs.

CHAPTER 5

ALL CREATURES GREAT AND SMALL (AND COLD)

Where we encounter strange life in the Frozen North; running
with dire wolves; the finest mounts Westeros has to offer;
how cold makes things huge; giant-killing with Sir Isaac
Newton and JBS Haldane; the truth behind zombification.

~ Mean Wolf! ~

The World of Ice and Fire has gone to the dogs. Actually, to be more
accurate, it has gone to the wolves: the dire wolves. This is especially

true once you get north of Castle Black, beyond the Wall, where they're everywhere. Worse still, these ferocious creatures are about the size of a small horse, and would think nothing of ripping off your arm and then beating you to death with the bloody end. The only good news is that they tend to stay in the furthest reaches of the North in the Lands of Always Winter, with Theon Greyjoy noting that the dire wolves haven't been seen south of the Wall for hundreds of years.

The Stark children, Sansa and Bran, Rob and Arya, Rickon and Jon Snow, find a litter of orphaned dire wolf puppies and are desperate to keep them – like all children everywhere when confronted with the cute, mewling faces of slavering ferocious death beasts in their juvenile form. Their father Ned gives them a lecture worthy of any parent in *Pets R Us* along the lines of 'Ok then but you'll have to walk them yourselves *even if it's raining*'. And it rains a lot in the North. Let us remember, Ned Stark is not a man known for his wise decisions, but this may be one of his better ones.

Dire wolves (*canis dirus* – Latin for 'fearsome dog') are unfortunately extinct today, but they were once real and could be found roaming the grasslands and forests of America, between 10,000 and 240,000 years ago. We know this because the fossilised remains of a staggering 3,000 dire wolves have so far been found in tar pits across California. The pits were actually pools of black, sticky asphalt that were created when crude oil seeped up from deep inside the Earth through a crack in its surface. In the last Ice Age, hundreds of these large black pits acted like vast sheets of flypaper, quickly immobilising any creature unlucky enough to venture their way.

Let's turn back the hands of time and watch the pit in action. Imagine a woolly mammoth, out for an afternoon amble. Seeing a pool of water that has formed atop the tar, the large beast inadvertently wanders into the pit and is somewhat surprised to see its hairy feet sinking slowly and inexorably into the black goo. Panicking, the mammoth raises its trunk and trumpets in distress. Sensing an easy kill, a nearby dire wolf hears the cry for help and rushes in. Moments later, both animals find themselves stuck in the tar, both unwilling participants in what researchers now refer to as an 'entrapment event'. Over time, the two animals start to feel strangely comfortable in one another's company, but slowly die from a heady mixture of malnutrition and boredom.

Thanks to geologists, we know that the dire wolves of our past were roughly the same size as our modern grey wolves, though broader and more muscular – weighing in at around 25% more – and with shorter legs (which meant they were probably slower to cover distances). If a dire wolf did catch you, and you found yourself in the unlucky position of being between its jaws, the bite force of this beast was pretty hefty; it would chow down on you with 129% of the force of a modern grey wolf bite (which itself is no playful nip).

Fossil records from the tar pits also show that the dire wolf apparently lived in large groups, often took down big prey, and tended towards a facial expression consistent with 'I can't believe I did that. God, I'm bored'. Unlike grey wolves, they would have steered clear of the snow and ice of the oncoming winter in *Game of Thrones*, as they preferred temperate regions. The most recent research has revealed that the wolves died out just over 10,000 years ago, and that their demise was possibly due to stiff competition from smaller animals, with coyotes proving especially efficient at picking off their shared prey.

So there you have it. Dire wolves did indeed exist and were fierce killers. However, unlike the wolves that populate the World of Ice and Fire, they weren't all-powerful and were eventually outsmarted by a group of well-organised and wily coyotes.

⤳ They Might be Giants? ⤳

Mysteriously large creatures emerging from the icy darkness to hunt, pale, spindle-legged spiders bigger than your face. The creatures of Samwell Tarly's White Walker nightmare. Welcome to . . . no, not the fictional Lands of Always Winter, but our own factual-but-strange world of polar gigantism!

As the realms to the north of the Wall may be alien and unknown, even frightening, to the inhabitants of Westoros, so the natural world of our own coldest places on Earth is strange to us. (Though it's worth pointing out here that we humans have been to our North Pole and, as far we know, the only weird white-bearded guy with strange powers hanging around there is Santa Claus, not a White Walker.)

When 19th century explorers began venturing forth with scientific expeditions to the North and South poles, the creatures they encountered gave them chills, in more ways than one. For a thousand years or more, Europeans had been venturing into the Arctic Circle to hunt for whales, the leviathan of the deep. They had met curious critters that seem fantastical, like the narwhal – the unicorn of the seas with its single twisted horn growing six feet long. But even when you're a hardy Victorian polar explorer, with glinting frost in your beard and partially digested, sadly loyal husky dog in your belly, nothing can prepare you for the sight of a squirming tide of *sea spiders the size of dinner plates*, all over the deck of your ship.

There are over a thousand different species of sea spiders, found all over the world, from the Caribbean to the Mediterranean. They tend to be tiny, sometimes just over a millimetre in diameter, but in the icy waters of the Antarctic they grow to over *90 centimetres*. And gigantism isn't just confined to spiders; polar bears are only found in the Arctic and are on average noticeably larger, and fiercer, than their browner, woodland-ier cousins like the Kodiak bear.

So what's going on?

Well, the giant spiders could be benefiting from higher concentrations of oxygen in the water and air, or better quality foodstuffs in the extreme north and south (these theories are still being debated). Or they could be a result of 'Bergmann's Rule', as posited by the 19th century German biologist Carl Bergmann. This is the principle that states that body mass increases in cold climates, and smaller size species are found in warmer regions. If we look back through evolution, it is true that megafauna (big animals) flourished at times when the planet was colder; and when the Earth heated up, our significantly smaller ancestors, the mammals, got their turn. Bergmann believed that it's all down to larger animals having a lower surface area to volume ratio, so they lose less body heat per unit of mass, and therefore stay warmer in cold climates. Warmer climates impose the opposite problem: body heat generated by metabolism needs to be dissipated quickly rather than stored within.

In *Game of Thrones*, of course, the tallest people we encounter are a race of giants who live north of the Wall, up in the coldest part of the Known World. Could polar gigantism also work on humans? Well . . . yes and no. Sadly there is no race of ancient giants hiding at the snowy wastes of our Poles, North or South. Instead, the people who live close to the North Pole are actually shorter, stockier and of heavier build than their counterparts closer to the equator. This fits with Bergmann's findings: someone of a stocky build will have a lower surface area compared to longer-limbed humans in hotter climates, which means less heat loss when it matters.

The human race is generally getting taller. We've grown on average four inches more since World War I ended. But the tallest nation on Earth is to be found in distinctly un-chilly Holland, with an average height of 168.7 cm (5 feet 6.4 inches) for women and 184.8 cm (6 feet 0.8 inches) for men. Researcher Dr Stulp, from the London School of Hygiene and Tropical Medicine, believes this is due to natural selection. It would seem that in Holland tall men have more children than shorter men, and are thus passing on their tallness as a heritable trait to their children. Will this go on and on, leading to a race of Flemish gargantuans, delicately stepping over windmills while munching on their boulders of Edam? (No offence, Holland.) Sadly not, as there seem to be limits to human height, beyond which the disadvantages definitely outweigh the advantages. Nevertheless, Gert Stulp intriguingly points to fossil evidence of Neanderthals discovered in Europe.

'We tend to compare ourselves to our historical counterparts in written history – 200 years ago – when we were much shorter,' he explains. 'But if we look at Palaeolithic records – 2 million or so years ago – we do find fossils that reach heights of 190cm and that is much taller than many of the individuals in populations now. That suggests that many populations haven't reached their evolutionary peak height . . . but it is difficult to say.'

(There was also a rather appealing theory doing the rounds for a while that Neanderthals' *noses* may have grown rather large due to adapting to the cold, but very sadly this seems to be untrue.)

⁓ Sir Isaac Newton, Giant Killer ⁓

Giants like Wun Wun, who live in the extreme north of Westeros, are generally about 10 to 12 feet (3 to 3½ metres) tall, though they can grow larger in some cases. In the books they are basically double the height of a man, in the TV show even taller. They are fearsomely strong and ride mammoths as steeds. So could they exist in real life? Or is there more than a touch of magic to them?

To create one – you just need a scaled-up, really tall human, right? Easy! Well, no. The 'scaling up' bit is exactly where we run into difficulties in the real world, thanks to the laws of maths and physics, which govern how changes in size affect things like movement and body mass.

Let's begin with what's known as the square-cube law, a mathematical principle that demonstrates why sizing up is a bit trickier than it first seems. Leaving aside the awkward, irregular and frankly confusing shape of a human being, let's begin by doubling the size of a cube. If you measure the edges of a cube and then create a new cube with edges double the length of the original, a strange thing happens. Yes, sure, the *size* of the cube *doubles* just as you'd expect, but its surface area doesn't double along with it; instead it increases *fourfold*, and the volume of our cube increases *eightfold*. (You could make a cube out of scrap paper, and then double it to see this in action, for a bonus *Ad Break Science* experiment.)

So, how does this apply to living creatures? Well, it's not an exact science as we're not dealing with regular shapes; plus, we'd want our giants to be strong and able to easily stand upright and move around, yes? Here we must turn to the mighty Sir Isaac Newton and factor in his Second Law of Motion, which states that force = mass x acceleration.

If we double the size of a human, while keeping them still roughly human-shaped, we end up with around four times the muscle power (so far so good), but that muscle power will need to move approximately eight times the mass. So this scaled-up creature is going to be only around half as strong and agile as a regular-sized human. Which means, for a start, our giant's heart and lungs would find it hard going keeping up with an army of regular-sized wildlings.

Ah, another great plan foiled by maths! But we run into this exact same problem all over the place with non-living things too. For example, it's an issue in engineering – from steam trains to rocket engines, inventors and innovators have had to grapple with the square-cubed law when trying to improve and create bigger and better machines. And it's also the reason skyscrapers can only get so big.

It's a little ironic when we think of Sir Isaac's famous and apparently modest line about his achievements: 'If I have seen further it is by standing on the shoulders of giants.' Yes, thank you, Newton, that'd be the giants who you've just helped prove couldn't really exist then . . .

Maths & Dragons

Also falling foul of the square cube rule are flying creatures like Dany's dragons. Dragons, as we definitely know by now, are magic. But if they weren't, they would very quickly start running into problems, as they regularly double in size, we're told. Each time they double in size, their wings will just about quadruple (yay – huge wings!) but the bulk they're trying to heft into the skies is going up near eightfold, so more and more lift is needed, meaning the dragon's flying ability would halve. So, pretty soon Drogon et al are going to find it increasingly hard to fly themselves, let alone carrying Daenerys Stormborn The Unburnt and her increasingly long-winded titles.

The Scientific Secret of the Giant's Causeway

Over in the Seven Kingdoms, giants are said to have had a large (very large) hand in building the Wall. In our world, the most famous feat of engineering commonly attributed to those of taller than average stature is the Giant's Causeway. Stretching out into the sea towards Scotland from the north-east coast of Northern Ireland, the Causeway is 'paved' with thousands of interlocking six-sided columns of basalt rock. By a strange quirk of geographical fate, this same breathtakingly beautiful stretch of coastline serves as a major location for *Game of Thrones*, with the area around the Causeway forming the backdrop to several scenes from the Iron Islands, home of Theon and the other Greyjoys.

How did the Causeway come to exist? According to folklore, an Irish giant called Finn MacCool was challenged to fight by a Scottish giant named Benandonner. There wasn't the big money in professional fighting in those days, and in the absence of a promoter to send round a limo, Finn himself was forced to build a rocky road so that the two giants could meet for their epic super-enormous-weight bout. However, as with so many large-scale multi-national engineering projects conceived in the heat of the moment and fuelled by hostility, regret soon set in (see also the Channel Tunnel).

As Finn is close to completing the Causeway he gets a better view of Benandonner, and realises there's no way he wants to fight the guy – Benandonner is considerably bigger than he looked back when they were in different countries with no convenient land bridge. So, his only option is to outsmart the mighty Scot. Enter Finn's helpful wife Oonagh, who dresses her husband as a baby and tucks him up in a cradle (we are not told whether this is a regular feature of the MacCools' marital life). When Benandonner sees the hulking infant he becomes fearful of the kind of giant who could father such a vast child. Tail between his giant legs, Benandonner runs back to Scotland, destroying the Causeway behind him so he can't be followed.

So far, so fantastical.

The site was first revealed to the world in 1693, by an Irish politician, Sir Richard Bulkeley, who presented a paper on the Causeway to the esteemed members of the Royal Society in London. It has fascinated geologists ever since.

The phenomenon of distinctive hexagonal columns is seen at various – usually supernaturally-named – sites around the world, including California's Devil's Postpile. Years of speculation and experimentation have resulted in scientists now firmly believing that the regularly shaped basalt columns are of volcanic origin. According to this theory, around 50 to 60 million years ago (during what's known as the Paleogene Period), the area around the Causeway was an intensely volcanic place covered with molten basalt lava. As the lava cooled it contracted, and eventually began to crack, like mud drying out on a hot day. The shrinkage caused stresses that then fractured the rock. But the strange, uncanny organisation of these cracks, or 'joints' as geologists call them, makes hexagonal shapes in a variety of sizes that appear very well ordered. And it is this that has led to the supernatural explanations.

MAKE YOUR OWN MINIATURE GIANT'S CAUSEWAY

SO why the six-sided causeway pattern, with different sizes of 'paving stone'? A few years ago physicists from the University of Toronto solved the riddle of the Causeway using materials you can find in your kitchen. In fact, you can carry out your very own version of their (literally) ground-breaking experiment right now, in the time it takes for a *Game of Thrones* ad break.

For this experiment you will need:
  A coffee cup
  A cup of cornflour
  A cup of water
  A bright light that you can leave switched on for a while

Begin by half filling the coffee cup with cornflour and then adding the same amount of water. Mix the cornflour with the water, and then put the cup under a bright light for seven days (the time between two episodes of *Game of Thrones*, conveniently enough).

After a week, the mixture will be dry. Carefully remove it from the coffee cup and break it apart. You'll see that the interior of the mixture has broken up into miniature hexagonal columns that are only a few millimetres wide. Congratulations! You have just created your very own Miniature Giant's Causeway.

The physicists who created the experiment studied the mixtures produced under many different conditions, as well as lava patterns around the world, and concluded that the size of the columns depends on the speed that the water (in the experiment) moves through the mixture. They wanted to know more about how the size of the hexagonal columns was determined – these structures vary from a few centimetres to a few metres across. Volcanic lava 'cracks' into columns much more slowly than cornflour mixture, explaining why the hexagonal 'paving' columns we see on the Giant's Causeway are so much bigger.

And that's how to explain some of the most spectacular scenery in *Game of Thrones*, with a simple cup of cornflour!

On Being the Right Size

When it comes to being a large creature, being on land or in the air is no match for the buoyancy of the ocean, which helps no end in taking the strain off internal organs, and so forth. It doesn't really matter so much about your *mass* in the sea, it's your density that counts – hence of course a small dense object like a marble will sink, whereas a huge sphere made of polystyrene will float. There's a reason the biggest living beings on Earth are in the sea. Animals are basically made of water, so the sea is the place to be, to escape the effects of gravity. So yes, a giant merman/mermaid? Perfectly plausible (see page 232 for more merfolk in action).

Wherever you are, is there a 'right' size to be? How do animals' relative sizes affect their lives? Why don't you get really huge hulking mice, or really tiny elephants?

The question of appropriately large or small scale animals fascinated Indian evolutionary biologist JBS Haldane, who pondered the issue in his intriguing essay 'On Being the Right Size'. Haldane begins by seeing off the traditional giants of children's fairy tales, as well as more modern monsters such as King Kong or Godzilla. He posits that the storybook giants he remembers from his childhood – 10 times the height of their human adversaries – would have weighed a thousand times as much as a regular human, and seeing as human thigh bones break under about

10 times the regular human weight, they would have snapped their legs each time they took a step. (Haldane speculates that this is why said giants were often pictured sitting down, but confesses that he has somewhat lost respect for Jack the Giant Killer's prowess. Young Jack could've finished off the 'fee fi fo fumming' f*cker at the top of the beanstalk by asking him to stand up and say that.)

There are advantages then to being small. 'You can drop a mouse down a thousand-yard mine shaft; and, on arriving at the bottom, it gets a slight shock and walks away, provided that the ground is fairly soft. A rat is killed, a man is broken, a horse splashes,' says Haldane (who is beginning to sound a lot like Ramsay Bolton). But lest we fear psychopathy, Haldane explains his real interest – how the forces that buckle giants save the little mouse.

Discharged down into Haldane's thought experiment mine shaft, your destiny upon arrival at your final destination will be very different depending on your physical form during this nightmare descent. To begin with, the mouse's tale. If you're a mouse, you have a larger surface area in relation to your volume – and therefore your mass – than the horse waiting nervously behind you. So as a mouse you'll fall more slowly, for a start.

'Divide an animal's length, breadth, and height each by ten; its weight is reduced to a thousandth, but its surface only to a hundredth,' states Haldane, employing our cube rule from above, this time in reverse. So, as a mouse, your physical resistance to falling is relatively ten times greater than the driving force pulling you down. Because of its size, the unfortunate horse has much more kinetic energy than the mouse so, when our (imaginary) horse meets the bottom of the (imaginary) mine, this energy has to go somewhere; it has to be 'dissipated' a physicist would say. So yes, it's dissipated by – essentially – exploding everywhere.

If you really really must recreate this experiment drop a grape, then a watermelon out of a first floor window *after* having checked no-one is around for at least a radius of 1 mile and you can clear up the watermelon and not waste it entirely!

⤙ 'Dark Wings, Dark Words' ⤙

Message-carrying crows and ravens flock through the worlds of fantasy literature, delivering vital missives and generally moving along plots nicely. The World of Ice and Fire is no exception, of course, with the

Westorosi postal system entirely reliant on these feather-based couriers to fly messages from castle to castle. Indeed, throughout the show, pivotal news arrives via ravens, often received with foreboding: 'dark wings, dark words', as Ned Stark memorably mutters.

The ravens are cared for, trained and dispatched by the Maesters, the post-master-scholar-healer-scientists of the Seven Kingdoms. In ancient times, the birds could talk, but this knowledge has been lost and now they carry written-down messages attached to their legs. In principle, ravens in our world could be used to convey verbal messages. While possibly not up to learning whole missives, they are excellent mimics who can learn to talk better than some parrots. Watch, for instance, YouTube sensation and talking raven Mischief demonstrating his ability to say 'Hello' in a rather stern, deep voice and 'Hi' in the manner of an anime schoolgirl.

I digress.

Ravens may never have actually carried messages for us in our world, but the idea of them as messengers has deep roots in popular culture. Myths from Tibet to Ireland have seen the raven as a messenger for the gods. The Norse warrior god, Odin, had two ravens – Hugin (thought) and Munin (memory) – who flew all over the world, collecting information and reporting back to Odin every night with their news. According to the Icelandic sagas, Odin also spent quite a while fretting about when they would actually arrive with their deliveries, muttering 'I am worried about Hugin yet more about

Munin', before apparently adding 'Thorsday he was late, then on Freya's day he came not at all – but I found a card of crimson and white declaring there was "something for me" and that he had returned "while I was out", though I had not left Aasgard these long hours and the post office of Midgard to which he referred me is only open 'til 1pm tomorrow . . .'

In our world, pigeons are the ones entrusted with carrying information. For years, the puzzle of how homing pigeons found their way home went unsolved. Ten years of international study by animal behaviourists at Oxford University, however, provided the answer. Apparently they simply follow our roads and highways! In 2004 Professor Tim Guildford explained the team's surprising findings to the world's media: 'It really has knocked our research team sideways to find that pigeons ignore their inbuilt animal instincts and follow the road system . . . When they do follow a road, it's so obvious. We followed some which flew up the Oxford bypass and even turned off at particular junctions. It's very human-like.'

∽ Getting to Crow You, ∾
Getting to Crow All About You

Despite their tiny bird brains, there is plenty of evidence to suggest that ravens and crows are as smart as chimpanzees and dolphins, and surprisingly sophisticated when it comes to their behaviour. Seemingly they are also capable of feeling empathy, and both recognise the concept of death and fear it.

But how did researchers discover this, I hear you ask? It's a good question. They asked a flock of crows to watch the 'The Rains of Castamere', aka the Red Wedding episode, and counted the number of times the birds lifted their wings to cover their eyes in horror. Just kidding. Actually, the researchers noticed that crows behave 'unusually' around death. When a crow passes on to the great flock in the sky, members of its group have been observed crowding around the corpse, squawking loudly, which intrigued researchers and got them wondering if the birds were taking part in a funeral for their dead co-crow.

Scientists at the University of Washington in Seattle set up a crow-death experiment. Lead researcher Kaeli Swift made a good impression on a flock of crows by feeding them treats. However, capitalising on the knowledge that crows do not forget a threatening face, Swift was soon joined by a second researcher: this time an apparently dangerous individual.

As BBC Earth reported, 'To prevent any real life harassment from crows, the face they used was not a real one but a rather realistic latex mask covering their real face.' So there you go. In the past, in the name of acquiring new knowledge, great men and women of science have exposed themselves to dangerous levels of radiation (Marie Curie), studied optics by impaling their eye on a needle (Isaac Newton) and even kept their baby daughter in an experimental box (BF Skinner – she quite liked it, apparently). But here, finally, we see a limit being reached. No-one wants to be disliked by crows. No-one is willing to become a hate figure for the feathered community, a crow-loathed miscreant unable to step outside the lab for fear of being cawed at and heckled by a testy corvid mob. (We've all seen Hitchcock's *The Birds*, right?)

This second researcher would 'be holding a dead crow, not violently,' Swift noted matter-of-factly, 'not re-enacting a death scene, just holding it like they were picking it up to throw it in rubbish, *palms outstretched like you might hold a plate of hors d'oeuvre.*' [My italics. As someone who is a fan of hors d'oeuvre, I'm not quite sure what to make of all this, other than I don't think I would ask Swift to pass round the nibbles at parties.]

Previous research has shown that crows not only remember a threatening face, they share that knowledge within their community, so that the individual is remembered and scolded by the crows, even after a gap of several years. Young crows, it seems, are even taught to recognise and scold the 'villain' by their parents.

The crows did not like the bad guy in Swift's experiment one bit; they began scolding and mobbing the masked researcher in a display of aggression, which also seemed to serve as a 'learning to recognise' experience for their flock. They also avoided Swift's food. Even when the masked researcher returned the next day without the dead crow, they avoided him and the food that Swift had brought them, giving a strong indication that they recognised and feared death, according to the scientists. (Though when the experiment was repeated with a dead pigeon the crows seemed noticeably less bothered – indicating that they recognised the threat specifically to their own kind. Or that nobody – not even other birds – really likes pigeons. Even if they are good at following highways.)

In the absence of warging, science appears to bring us tantalisingly close to seeing the world through the eyes of animals. But as laypeople, we are still prone to misinterpreting our corvid friends. Ravens are one of the few animals that we believe reliably use gestures – with beak or wing – to draw the attention of fellow ravens to points of interest. And they appear capable of communicating with humans, too.

In his book *Mind of the Raven*, Professor Bernd Heinrich recounts a curious encounter between a human, a raven and a cougar. In rural

Colorado, a woman was working outside her cabin alone when she became aware of what seemed like the increasingly urgent chattering of a raven. When the bird landed on a rock nearby she looked about her and came eye-to-eye with a cougar, poised to spring and attack her.

How would you interpret this encounter? The woman (who escaped unscathed) believes that the raven was trying to warn her that her life was in danger. Heinrich, who has spent many years studying the behaviour and culture of the birds, believes rather that the raven was identifying her as a dining opportunity for itself and the cougar, and warmly encouraging the big cat to stop wasting time and *get on with it*. It's not for nothing that the collective term for a flock of ravens is 'an unkindness' or 'a conspiracy'. The carrion birds were traditionally regarded as sinister; like other corvids, they would flock to battlefields to pick over the dead and dying, and of course the fourth book in the *A Song of Fire and Ice* series deals with the terrible aftermath of the War of the Five Kings in *A Feast for Crows*.

In the wild, ravens will certainly team up with other animals, including humans, for a mutually beneficial arrangement. Wolves and ravens, for example, will pair up to hunt. When wolves were reintroduced into Yellowstone National Park in the 1940s this complex attachment was observed re-establishing itself very quickly – the wolf pack was seen hunting with ravens flocking close behind. That both these creatures, wolves and ravens, have a special place in our mythology and stories, and that their behaviour still intrigues us, hints at a time when we were closer to the animal world, possibly when ravens were indeed bringing 'messages' of a kind.

Is it entirely fanciful that our first hunter-gatherer forebears would have recognised and followed the swooping of ravens to a kill by wolves as a way of finding food when times were hard, and that all three creatures hunted together? When you're hungry enough, the cawing of ravens sighting meat really could seem like the most precious message from the gods.

∽ If We Could Ride Any Animal ∾

Getting from place to place in the Seven Kingdoms involves a variety of four-legged transportation. Usually, as you might imagine, it's fastest to ride a horse. But some rather more exotic creatures have been pressed into service in Westeros.

Dany, like her Targaryen ancestors, rides a dragon – initially with varying degrees of success. (She and Drogon end up making a good team, but finding dragon riders who can control the fire-made-flesh of Drogon's siblings, Rhaegal and Viserion, seems crucial for the conquering to come.) In the books we don't hear much about the northern, island-dwelling Skagosi – they're a brutal people, legends say they are cannibals – but in an unexpectedly whimsical move, they trot about on unicorns. There's Cold Hands, a kindly, enigmatic book character from north of the Wall who helps Gilly and Sam as well as Bran and the Reeds. Notably, he rocks up on the back of a large elk. And if we were to journey across the World of Ice and Fire to the eastern continent of Essos, we'd encounter the nomadic raiders Jogos Nhai, trekking across the great open plains astride their black-and-white striped equines, zorses.

So, can any animal with suitable human-sized space on its back be ridden? And in our world, why do we still ride horses, rather than, say, zebras or zorses?

Well, it's not black and white . . .

We humans have been domesticating our fellow animals for about 20,000 years. And in that time, we've figured out there are certain common factors that make a creature comfortable to be around. They can't be too picky about what they eat, they must be ok with breeding and bringing up their young quite swiftly in captivity, and they must

143

have a hierarchy, so that humans can replace the top dog (or horse, or dragon). They also need to be quite calm and relatively sweet tempered. And this is where, for example, zebras don't really earn their stripes.

Zebras – like the zorses of the Jogos Nhai – are highly unpredictable and prone to biting and kicking if they don't like the look of you. However, at least one man in our world saw that as a challenge to be overcome. At the end of the Victorian era, if you found yourself in the smart part of London near Buckingham Palace, you might have witnessed the memorable sight of financier-turned-zoologist the second Baron Rothschild, riding along the road in a carriage pulled by four zebras. There are other reports, too, of zebras being ridden like horses. It's just you need to be quite brave.

When Jon Snow ventures north of the Wall, he's bewildered not only by giants twice the height of regular men, but also their mounts – mammoths. Sadly, while we have no cave paintings depicting our ancestors from prehistoric times triumphantly posing atop shaggy pachyderms, historians believe that elephants have been giving humans a lift for at least 6,000 years. They carried the invading Romans across the Thames and into Britain in AD45 (we didn't like that one bit) and in World War II, the allies used elephants for transportation in tropical regions, where the modern vehicles of the time couldn't traverse. Today, in peace time, elephants carry tourists around the Amber Fort in Jaipur, India.

Elephants do contradict our guidelines for domestication and as they aren't easy to breed in captivity have been taken from the wild and trained (and ridden) by humans for millennia. We have a unique insight into this process from the ancient Indian text the Arthashastra – a Sanskrit treatise written around 2,000 years ago, which offers guidance to rulers on government and politics of state. The Arthashastra advises that 'summer is the time to catch elephants', and recommends seeking out a wild elephant that's around 20 years old (presumably they are particularly easily enticed at that age – like dangling an unpaid internship in front of a newly graduated student). It goes to on to provide details of how the beasts should be trained, cared for, stabled, rested, and exercised in as kindly a way as possible – with suggested penalties for any human who doesn't treat the elephants well. It also recommends a kind of 'pachyderm pension' from the treasury be set aside to look after elephants who can no longer work or be ridden to allow them to live out a peaceful, pleasant old age (not unlike our State Pension). This is perhaps why according to reports in late 2015, the Indian Supreme Court was considering a ban on elephant rides as entertainment, due to concerns about these peaceful creatures being kept in cruel, inadequate conditions.

While the sight of giants riding on mammoths may have amazed Jon Snow, there are more extraordinary sights north of the Wall. The Others, it seems, have their own unique way of navigating the Lands of Always Winter. While among the wildlings north of Castle Black, Samwell hears the horn of the Night's Watch blown three times – three long chilling blasts which mean the White Walkers are nearby. He remembers the stories of monsters that made him shiver and squeak as a child, the descriptions of the approaching blood hungry predators *riding their giant ice-spiders*.

Sweet dreams . . .

THERE'S NO BUSINESS LIKE SNOW BUSINESS

AFTER all that talk of White Walkers, do you, er, wanna build a snowman?

At a time when interest rates are more likely to be frozen than the Thames, snow can be surprisingly hard to come across. But worry not, help is at hand. For years, scientists have toiled away in their laboratories creating the perfect formula for fake snow. And the even better news is that you can enjoy this magical white powder in your own home. It's time for Ad Break Science!

For this experiment you will need:

A disposable nappy (or diaper, for the Americans) (the super-dry type that contains liquid-absorbing crystals at the 'business end')

A pinch of glitter (optional, but looks fabulous)

A cup of water

A bowl

Cut open the nappy and remove the treasure trove of magic crystals. Place them in the bowl and gradually add water. Just add a little water at a time and mix everything thoroughly until it all becomes suitably snowy. Add a sprinkle of glitter and you have created your very own slice of winter wonderland.

The special 'crystals' inside the nappy are made of a super-absorbing polymer called sodium polyacrylate, also called 'slush powder'. When the crystals encounter water, the sodium atoms whizz around and swap places with the water molecules, expanding rather than dissolving and creating a wonderful substance that looks and behaves just like snow.

If you are sans nappy (or if the state of the nappy would result in the classic 'yellow snow') don't worry! Instead, you can use some bicarbonate of soda (baking soda) and some snowy-white coloured hair conditioner. Mix at a ratio of 3 cups of bicarbonate of soda to half a cup of hair conditioner. Finally, sprinkle in a pinch of glitter for added sparkle.

Conditioner is acidic, which is why it settles your hair down after you've cleansed it with shampoo (which is alkaline). Bicarbonate of soda reacts with the conditioner to produce carbon dioxide. The resulting concoction has the fluffy-graininess of snow (thus the phrase 'I just wash and snow'). As an aside, when the nuclear bomb was being developed in the 1940s, researchers discovered that bicarbonate of soda was the perfect thing for washing uranium out of clothes, which is both helpful in the event of a future nuclear war and may earn you an extra point at the next pub quiz/trivia night.

SHAMPOO DU TARGARYENNE

So, however you make it, roll up a small ball of fake snow, pop it on top of a slightly larger fake snow ball, add a few black feathers, some dark hair, a swoonily handsome face, a black cloak, a little sword maybe? . . .

Ah, you know nothing, Jon Snowman!

Everything here is non-toxic and found around the home in everyday products. But do avoid getting the fake snow in your eyes/mouth/nose/orifices generally (!) just to be on the safe side, as it may cause irritation for some. And when you've finished with your snowman, put it in the bin/trash rather than down the sink.

~ The Strange Truth ~
about Zombification

An uncannily cold blue gaze shines down from the pale, impossibly wrinkled regal face and falls upon the baby. The gathered shambling hordes are silent, sensing the majestic importance of this moment. As the crowned elder reaches down to touch the child's cheek, a transfer of ancient power is complete . . .

No, this is not a royal family christening but the chilling scene in Season 5 in which we finally discover the icy fate of Craster's male children. Torn from their mother's arms and abandoned in the woods by their deeply deadbeat dad, the baby boys are gathered up by the White Walkers, carried away to the Lands of Always Winter and transformed into chill-eyed beings of pure coldness by their new 'family'.

The White Walkers (also known as 'Others') were originally a source of great mystery. What did they want? Were they really as evil and destructive as they seemed? And were they available for children's parties?

The chilling power of these cold shadowy creatures is clearly A Big Deal because we encounter it in the opening scene of both the book and the TV series. Initially the macabre horror of what we're seeing is mysterious. Three members of the Night's Watch stumble across the bodies of men, women and children, butchered in the snow in a clearing in the forest. The next thing, all their bodies have vanished. And then, in helpless horror, we watch as a child, previously lifeless and impaled through the heart on the branch of a tree, stands and coolly surveys the Night's Watch with unnaturally piercing, bright blue eyes.

As the story progresses, we learn more about the supernatural White Walkers. They're an ancient enemy, demons of ice and cold, the only enemy that matters, says Stannis. They haven't been seen for millennia and had receded into legend. But as we enter the World of Ice and Fire, they are reappearing, and bearing down on their human prey with shocking violence.

At the Battle of Hardhome, we witness their terrifying ability to violently mutilate human beings and then resurrect them as wights – mindless thralls who will do the White Walker's savage bidding (relentlessly, until every last part of them is used up in the macabre service of their masters). As we see to dramatic, jaw-dropping effect, these wights eventually form a huge 'zombie' army under the control of the Others – ruthlessly killing their fellow humans as quickly as their overlords can resurrect the fatalities and swell the ranks of their undead army still further.

The White Walkers ride gruesome living-dead zombie horses, too – all torn sinews and bloodied-spittle nostrils. So we can assume that their power also reanimates animals.

Necromancy, bringing the dead back to life and setting them up to no good, is something we read about in the Bible, and hear about from time to time throughout history. Yet no-one modern and sensible seriously thinks it's real. Right?

Wrong!

In 1985, Harvard ethnobotanist Wade Davis published *The Serpent and the Rainbow*, his account of 'an astonishing journey into the secret society of Haitian voodoo, zombis [sic] and magic'. Davis argued persuasively that zombies were both possible and real. He believed that cultural forces and beliefs framing the ingestion of special 'zombie powders' could produce some of the undead-esque behaviour recognisable to fans of George Romero movies.

According to his account, a Haitian shaman – called a *bokor* – would initially give the zombie to be a drug derived from pufferfish venom (tetrodotoxin), which would temporarily produce the 'effects of death' – paralysis, stiffness, 'the odour of rot' – without actually killing you. The proto-zombie could even be buried alive, so that he and others would become convinced he had experienced death, before he was 'resurrected' by the shaman. The zombie would then be rendered forgetful, delirious and suggestible with the administration of yet another potential toxin, this time a plant known as Datura (or more evocatively called 'Angel's Trumpets').

Davis secured samples of zombie powder, but as one of the powder's ingredients is bits of dead human tissue, he was heavily criticised for commissioning a grave robbery, to obtain the decomposed flesh of a recently buried child. In the end, his critics noted that many of his samples of zombie powder failed to contain tetrodotoxin, and even when small doses of the neurotoxin were present, it wouldn't produce the effects described in Davis's account. If we really want to get to the nasty truth about zombies, we have to turn to the peculiar relationship of spiders and wasps . . .

~ Zombie Spiders and Other ~ Body-Snatching Stories

From the White Walkers' ice spiders as 'big as hounds', to the walking-dead arachnids of our world, playing host to parasitic wasp larvae, unlucky spiders play a key role in our zombie story.

Boohoo, you may think, couldn't happen to a nicer creature.

But read on . . .

Some arachnids are continually engaged in a life and death battle with another type of malign creepy crawly: the parasitoid wasp. There are hundreds of thousands of types of wasp – each has its own way of being unpleasant. Here, we look to parasite wasps who build nests to rear their young *inside other living creatures*. Czech scientists Stanislav Korenko and Stano Pekár describe the effect of one wasp, *Zatypota percontatoria*, which hones in on the abdomen of a spider and injects its egg right in there. When the egg hatches the larva feeds on its host.

Initially, for the spider, it's business as usual – the host keeps spinning webs to catch flies. But when the larva is reaching maturity, the zombie mind-control kicks in. The spider slave stops its normal web-spinning pattern and begins spinning a different kind of web – one that is an ideal dream home for a wasp cocoon, with a platform to keep it off the ground and even a hood to shield it from the weather. For the wasp larva, the spider becomes architect and couturier in one. And when the specially commissioned web is ready, the larva bursts from the spider, killing it, and spins its own cocoon in the web.

Messing with a spider's web-making is only the tip of the zombie-wasp iceberg. Take the small, pretty and utterly terrifying jewel wasp, for example. This parasite is only a few millimetres in size, yet it builds a nest for its eggs inside a living cockroach by zombifying the far bigger, stronger host. The process is queasily fascinating to watch. The jewel wasp carefully selects its victim. It cannot overpower the cockroach, so instead it uses neurotoxins and 'mind control', the like of which humans cannot match, and even the White Walkers would admire (well, maybe not. It's hard to imagine them taking time out from their icy evil to watch a David Attenborough documentary).

The first sting the wasp administers to the cockroach makes the creature's front legs buckle so it is unable to escape; the second sting is delivered with a remarkable precision to the exact area of the cockroach's brain that seems to control its movement and (if we can speak of such a thing in relation to a cockroach) its free will. Scientists have actually developed a 'vaccine' that can 'de-programme' zombie roach if they can get to it at this point in the process, so there is some hope.

Now because the wasp is so tiny compared to the cumbersome cockroach, carrying her host is out of the question. Instead, the wasp bites down on the cockroach's antennae in order to lead it to the exact spot she wants it to go, 'like a dog on a leash'. (Some researchers have speculated this also enables the wasp to 'taste' how well her neurotoxins are taking effect in the cockroach's blood; it has yet to be proven, but whatever she's doing it's far more sophisticated and precise than anything a human brain surgeon could master).

Just like the wights and the White Walkers, the cockroach can still move of its own accord, it's just lost the will to *unless its wasp mistress leads it.* When they arrive together at the spot the wasp has decided on for her nursery, she digs a hole (the parasite wasp equivalent of painting the walls pink or blue) and then lays her egg on the underside of the cockroach, who also decides, hey, I like it here too, so it isn't going anywhere . . . ever again. Again, just like the wights, whose severed limbs will *still* serve the will of the White Walkers until they are completely destroyed, every last part of the cockroach is used and subjugated to the higher purpose of the wasp's young. The egg will hatch and the wasp larva will burrow headfirst into the cockroach's uncomplaining body. Hungry, the picky young larva delicately eats some of the spare 'stuff' in its host's thorax, thus ensuring the still zombified, unresisting creature is kept alive by leaving the cockroach's vital organs until last. (It delicately munches its way round them, like pizza crusts.) Then, one day, when the wasp larva has grown big enough and strong enough from all that delicious living cockroach, it bursts out of the cockroach's body, shakes its wings free and flies away. Finally, having served its purpose (as well as breakfast, lunch, and tea) the roach host dies.

Some parasites appear to possess utter ruthlessness alongside a wonderful sense of humour. A parasitic nematode infects ants and doesn't mess with their minds much at all; it just gives them an enormous red derriere. The affected ant gets spotted by a bird, who misses the joke somewhat and mistakes the ant's junk-filled trunk for a delicious berry. A moving berry, with legs, that's waddling along with a whole line of perfectly normal ants. Well, the bird is busy and if you're a hungry bird you don't look a gift berry in the . . . whatever. As it happens, when it comes to delicious berries, birds are always ready for this jelly. So, that's how nematode eggs get inside a bird, the bird then defecates them out from a great height – thus spreading its nematode eggs far and wide to entrap more hapless booty-ants.

These manipulations seem to have parallels with the complex interactions in the World of Ice and Fire between The Children of the Forest, the While Walkers and the Wights. As we discover in Season 6 of *Game of Thrones*, the Children of the Forest originally created the White Walkers thousands of years ago as a means of fighting back against the 'First Men' who had invaded their homelands. The Children captured some of these men and changed them from regular human beings into White Walkers, by injecting them in a particular way with dragonglass. (This is a little reminiscent of our earlier encounter with the jewel wasp, who injects the cockroach and turns it into the protector of its young – both the tiny wasp and The Children are trying to safeguard their futures by harnessing a stronger force to do their bidding.) However The Children seem to have lost control of the White Walkers at some point, and the White Walkers in turn begin creating their own 'zombie' servants, the Wights, over whom they exert much stronger control . . .

I talked to Dr Kelly Weinersmith, Huxley Faculty Fellow researching how parasites manipulate host behaviour in the BioSciences Department at Rice University, Texas, to understand more about this cunning behaviour in our world. She explains the behaviour and ultimate 'aims' of different types of parasite – 'subtle manipulators' versus 'all-out zombifiers'.

'Trophically transmitted parasites are parasites that need some or all of their hosts to be eaten by the next host in the life cycle in order to be transmitted. The nematode [who we met earlier] needs the ant's abdomen to be eaten by birds. These parasites need their hosts to die in a very particular way, since dying any other way would mean death for both the host and parasite.' Dr Weinersmith tells me that it's possible that the parasites in these cases are 'letting' their hosts generally behave normally, 'since natural selection has likely favoured hosts that behave in ways that tend to keep them alive. By altering just one or a few behaviours, the parasite may be able to increase the probability that the host survives until it has a chance to be eaten by just the right kind of predator.'

On the other hand – parasitoids are more likely to all-out zombify their victims. Like the White Walkers who ruthlessly control their zombie servants', the Wights every movement, parasitoids leave nothing to chance. Dr Weinersmith quotes one of her colleagues, Professor Shelley Adamo, who calls them 'evolution's neurobiologists' as they can do things human neuroscientists can't do (though in some cases this is definitely for the best). In a filmed talk for Smithsonian magazine, Dr Weinersmith gives a breathtaking example of the way a tiny fungus is able to control its ant-host, manoeuvring it to a precise height (25cm) on the North-North West side of a plant's leaf, finding its way to a leaf vein at solar noon. 'Wow!' exclaims an audience member. 'Yes – wow!' says Dr Weinersmith. 'If your mind isn't blown by how specific that is we can't be friends.' But to return our earlier wasp and cockroach example, the wasp controls the cockroach's behaviour so it can hide it away somewhere where the host won't get eaten by predators. The parasitoid's eggs then hatch in the cockroach and eat the insides of the cockroach. The parasitoid is going to kill the host outright at some point (e.g., by eating it alive). Dr Weinersmith speculates 'maybe the important evolutionary difference here is that the parasitoid controls when the host dies, whereas the trophically transmitted parasites need to hang on until the right predator happens to come around.'

While this manipulative behaviour is fascinating by itself, it also has surprising potential benefits for possible future research into new types of drugs for humans. As we've seen, parasites are hugely adept at manipulating their hosts by affecting their brains. They are able, for instance, to exert an influence across the blood-brain barrier (that's the membrane that exists as a protective measure against the passage of potentially harmful substances into the brain). A fish parasite that Dr Weinersmith studies appears to have a remarkably 'calming' effecting on its host – preventing it from stressing out at – well, the usual things that stress fish out. Could greater knowledge of how this 'calming' effect works one day lead to better anti-anxiety drugs for humans? Possibly! So it seems that, as Dr Weinersmith points out – rather than focusing on the evils of mind control and zombie horror stories, if we can learn from these highly skilled manipulators, one day they may lead us towards diminishing rather than increasing human suffering.

Parasites Start Gold Rush!

Bizarrely, not everyone is appalled by parasitic behaviour. Some of us really want in on the action. In recent years, a moth larva fungus has created worldwide demand for a commodity called yartsa gunbu, aka 'Himalayan Viagra' (which makes it sound like it might work a bit too well, but it's named for the dramatic location where it's found rather than the dramatic results when it's taken.) Yartsa gunbu is created when underground caterpillars become infected by a parasitic fungus, which mummifies the caterpillar and eventually bursts vertically out of its hapless host's head. (A signature parasite move, as it turns out.) It is used in traditional Chinese and Tibetan medicine to treat cancer and also as an aphrodisiac. A couple of years ago it was reported that just 500g of yartsa gunbu would fetch $13,000 (£8,100) in Lhasa, Tibet, and double this in Shanghai, China. It was also a recent suggested 'morning smoothie' ingredient on Gwyneth Paltrow's lifestyle website goop.com.

⌐ Secrets in the Ice ⌐

In our world, as in the Lands of Always Winter north of Westeros, the cold keeps all kinds of secrets. Scientists working in Antarctica are using finely balanced drills to bore as deep as three kilometres into glaciers and ice sheets, extracting rods of ice known as ice cores to catch a glimpse of undiscovered history. These ice cores have allowed us to look back as far as 800,000 years, and there is hope that it will one day be one million years or more.

When I speak to Dr Tamsin Edwards, lecturer in Environmental Sciences at the Open University, she's just back from the British

Antarctic Survey (BAS) labs. There's a photo of her on Twitter beaming with utter delight as she holds a tiny off-cut ice chip to her ear. The Survey have a bag of leftover ice core chips and their visitors can rub them between their fingers and listen. When the ice is exposed to a bit of human heat the bubbles of ancient atmosphere trapped inside it audibly fizz out. Gases that were caught in ice when woolly mammoths, Ancient Greeks or ruffed Elizabethans roamed the Earth, are freed into the air again after their long, cold imprisonment. ('It's *such* beautiful ice too – perfectly clear, apart from lots of tiny, identical bubbles,' Dr Edwards emphasises, with an ice connoisseur's eye.)

Scientists can analyse the composition of trapped air bubbles in the ice core to give us a record of Earth's climate history.

Glaciers contain layer upon layer of fallen impacted snow, and to a glaciologist the sizes of the snow layers are like the rings of a tree to a dendrochronologist. It is possible for them to tell us, for instance, how much snow fell in a particular year. And thanks to tiny amounts of gases, dust and particles of debris that get trapped when fluffy snow falls onto fluffy snow and freezes, they can tell us more about conditions at the time the ice formed.

Nearby vegetation contributed ancient pollen to the ice cores; climate scientists can even see which way the winds were blowing from this pollen and detect other distinct detritus that has gusted in from elsewhere in the world. Likewise, violent volcanic eruptions leave a layer of ashy dust and tiny shards of volcanic glass amid the ice crystals, telling tales of cataclysms before recorded human history began.

Dr Edwards shows me data she uses in teaching that measures dust with lead in it that blew onto ice sheets many years ago and remained frozen and trapped ever since. (If, like me, you are not particularly fond of, or adept at, housework you may be used to uncovering extensive dust-based revelations about the past of your own environment – probably a few minutes before you're expecting an important visitor to arrive.) But the level of historical detail revealed by this ice core is extraordinary.

Climate scientists can identify the period when humans discovered cupellation (a process that metal-workers have been using from about 500 B.C. to extract gold or silver from base metals). We can see there's a distinct rise in icy lead dust when the use of coinage became widespread, first in Ancient Greece and then across the Roman Empire, leading to a pre-Industrial Revolution peak – before a sudden fall when the Roman lead mines were exhausted, and the Roman Empire itself began to decline.

From those bubbles of gas trapped in Antarctic ice cores, scientists like Dr Edwards can analyse the levels of carbon dioxide (CO_2) and other greenhouse gases in the atmosphere, way back into the past, and how those concentrations were created in the first place. It turns out that carbon atoms are rather good storytellers with a great memory for times and places. For instance, by measuring the ratios of different types (or isotopes) of hydrogen atoms, scientists can see the difference between CO_2 that arrived from living things and CO_2 that arrived from, say, burning fossil fuels once the Industrial Revolution really got going in the 19th century. Each slice of the ice core tells a story about the world it came from, its climate conditions, and how much ice existed on the planet at that moment.

Putting this information together, climate scientists came up with the 'famous' saw-tooth shaped pattern of climate change which illustrates our Earth's long history of ice and fire. Ice ages come along every 100,000 years or so, with greenhouse gases rising and falling in regular patterns. But accelerating technological change, and a growth in population and consumption mean CO_2 emissions have rocketed in the past 100 years, causing potentially catastrophic consequences for the human race.

Inevitably, climate change plays a major role in *Game of Thrones*. As George RR Martin told *Al Jazeera America* in 2013, 'climate change . . . [is] ultimately a threat to the entire world. But people are using it as a political football instead . . . You'd think everybody would get together.

This is something that can wipe out possibly the human race. So I wanted to do an analogue not specifically to the modern-day thing but as a general thing with the structure of the book.'

In our world, politicians and rulers squabble and go to war, ignoring the more dangerous, all-consuming threat of melting polar ice caps. In the Seven Kingdoms, the different noble houses are too busy arguing amongst themselves to acknowledge the icy threat beyond the Wall, posed by the sinister White Walkers.

Back in the real world, climate scientists like Dr Edwards are eager to reach out to those on all sides of the global warming debate, to discuss the limitations on what we can know, the increasing complexity of models for prediction, and to assess the risks (and opportunities) of climate change. Whatever your views, it's difficult not to be captivated by the beauty of the Antarctic ice cores – the waxing and waning of great empires recorded in ancient dust, water and gas found frozen in time deep inside an uninhabited continent at the ends of the Earth.

CHAPTER 6

ANOTHER ICE BRICK IN THE WALL

Where we scale the Wall; learn how sound waves can destroy solid matter; and recount the sorry tale of the cat piano (a keyboard worthy of Joffrey Baratheon).

∽ Build a Wall! ∾

The answer to many problems in human history has been 'Build a wall'. Sometimes the fortification goes up to keep out bothersome

Geordies and Scots (Hadrian's Wall); sometimes it's to stop ice creatures riding around on dead horses, raising armies of zombies (Westeros's Wall). Sometimes it's a case of cunningly exploiting the new era of mistrust in an established political class with unrealistic but populist policies (Donald Trump's Wall).

In *Game of Thrones*, the Wall is 700 feet tall (think two Big Bens stacked on top of each other somehow), 8,000 years old, and made of magic and frozen water. No-one knows exactly how it was built – the ancestors of the Starks and giants are said to have been involved. All we know is that it is essential for dividing the North of the Starks and their allied Houses from the lawless realm of the Free Folk and the mysterious Lands of Always Winter.

THE WALL

It is a thing of rare beauty, even by the standards of fantasy engineering. (In the TV show, the set designers mix their fake Wall snow with sea salt to give it a sparkle suggestive of the magic that built it.) But if we take away magic, the force of gravity would definitely have something to say about that skyscraper tall feat of ice and its feet of clay (as it were). When *WIRED* magazine asked several engineers about its architectural potential, the prognosis was not good. Ice is a great defensive material, they said: it can withstand plenty of bombardment from the various medieval-style weapons used in *Game of Thrones*. However, even at sub-zero temperatures, a wall made of ice would begin to 'deform' under the pressure of its own colossal weight. The lowermost parts of the Wall would end up bulging outward as the uppermost parts pushed down, similar to how glaciers flow downhill. The way to counteract this would be to build the Wall on a gradient slope, but to keep a 700 foot wall upright it would need to be around 40 times that in width. So, less of a wall, more of an ice rink . . .

～◦ Sounding out ◦～

Even in the world of *Game of Thrones*, the Wall isn't presented as inde-structible. Ygritte tells Jon Snow about the Horn of Winter, a magical instrument which, if blown, will bring the Wall and all its frozen history tumbling down. The wildlings have long been seeking it, rifling through ancient bones in mouldering graves to find the ulti-mate weapon in their quest to push south into the kingdoms of Westeros. It is so powerful, Ygritte explains, that it can wake sleeping giants. (All over Westeros, everyone's clamouring to find a magic horn for one reason or another. The Ironborn leader Euron Greyjoy has one at least in the books that he believes will tame dragons and win him Dany and the Seven Kingdoms.)

But what's the real-world power of an almighty blast of sound?

There is a story in the Bible, much beloved by some young children, of Joshua and Jericho. Joshua was Moses' second-in-command, and the Bible's most famous warrior. He was also, it seems, a man with a keen belief in the power of a good horn blast. While besieging the Canaanite city of Jericho, he bid his priests to blow their shofars (a ram's horn that is blown like a trumpet, and still used in synagogues today, though to slightly less dramatic effect) and the rest of the Israelites to raise a great shout in order to bring down the defensive city walls, which they duly did.

Archaeologists have struggled to find evidence of Joshua's sonic coup in the historical city of Jericho. Nevertheless, in the 1990s, America's The Learning Channel decided the time was right for further investigation. Doubting the lung power of even the most highly motivated musical priests, the plan was to determine if some sort of noise generating device might have been at the disposal of Joshua & Co. They commissioned Wyle Laboratories in California to build a small brick wall and pit it against the largest loudspeakers available, turned up to 11. Wyle make a speaker called the WAS 3000 that omits a sound roughly 10,000 times the volume of regular home hi-fi speakers. After just six minutes, the brick wall capitulated and crumbled obligingly before the wall of sound. I don't think anyone's suggested that Wyle's speaker technology was available in the late Bronze Age when Joshua lived, but we've kind of proved something here, I guess . . .

And sound *is* powerful, we've all seen the films where an ill-timed shout or noise brings down a devastating avalanche of snow in the mountains. So could we achieve such results? Is sound a force that can become a weapon?

♫ DERRR
DUUURGH
DER DER DUHHH
DER DER DUHH ♫

As ever, the genius Kate Bush may have been on the money with her song 'Experiment IV'. In the music video, a sinister military commander (played by actor Peter Vaughan, who coincidentally also portrays Maester Aemon, friend to Jon and Samwell in *Game of Thrones*) secretly forces scientists to create a sound that could kill someone from a distance. There have long been rumours that such a weapon might be in development or possibly may even exist.

Infrasound is usually defined as any sound that is lower in frequency than 20 hertz (just below the limits of normal human hearing). It can be produced naturally from ocean waves and earthquakes, with the 1883 Krakatoa volcanic eruption producing infrasound that circled the globe several times.

In the 1960s NASA discovered that such sounds could disrupt respiration, produce headaches and kick off coughing. Strangely, the waves can also move small objects and even cause the flickering of a candle flame, causing some researchers to speculate that they may even be responsible for ghost sightings in a supposed haunted house. More menacingly, these low frequency sounds have also been investigated by the military as a possible source of acoustic weaponry. They are often referred to as the 'brown note' as the vibrations can allegedly affect the listener's bowels, causing them to defecate uncontrollably. Indeed, in 2000 an episode of *South Park* involved the inadvertent broadcasting of the note on national radio, causing millions to simultaneously empty their bowels.

Continuing US television's longstanding love affair with making sounds that can ruin absolutely anything, science show *MythBusters* then examined the idea by subjecting people to high levels of infrasound and discovered that the worst effect involved people feeling nauseous. In short, no need to worry about being acoustically fired upon just yet.

Concerto for Unhappy Cats

In the 17th century, German inventor and scholar Athanasius Kircher invented a number of wacky machines connected to the power of sound and harmony. In the 1600s he pondered building giant ear trumpets into the walls of palaces; he designed an automated head that could 'speak'. And, most terrifyingly of all, he designed the *katzenklavier* (a piano made out of cats). What?! Well, yes, it's a keyboard with seven to nine kittens held in cages, whose tails were attached to piano keys fitted with pointy mallets. Each unfortunate feline had a different (not entirely melodious) yowling tone and thus when a note was played the poor cat would react noisily to the tail-based-affront, resulting in a melody of meows.

Fortunately, Kircher never actually built a *katzenklavier*, and in spite of his eccentricities he's remembered as a great mind. The *Encyclopædia Britannica* hails him as a 'one-man intellectual clearing house', but maybe those *Britannica* guys are more dog people? In theory, he imagined it would provide a unique boost to 'melancholy princes', surprising them from their gloom and provoking 'laughter'. And possibly this would work if the prince in question had been Joffrey, but certainly not if it were his cat-loving brother and successor King Tommen, proud owner of the fine feline Ser Pounce (and bonus kittens Boots and Lady Whiskers).

By the way, if you've ever wondered who has the time to give their pets affectionate and silly names amidst a backdrop of war and mass killing, such as Westeros or the Middle Ages Medievalist Dr Kathleen Walker-Meikle from University College London has discovered that in the real world Gilbert was the most common given name for cats in England in the Middle Ages; an Irish monk wrote a poem about his beloved cat Pangur Bán ('Fair Fuller'); and we know a mouser called Mite stalked around Beaulieu Abbey in the 1300s. (Names deemed appropriate for dogs at the time included Troy, Bragge, and, delightfully, Nosewise.)

ᕽ Dragon Binding ᕽ

We know that animals can be made to make ungodly sounds, but a second question is how responsive are they to sounds sent their way? Recent research has shown that a great variety of animals – whales, elephants, squid, guinea fowl and rhinoceros – are sensitive to low-frequency sounds, and use sonar to migrate and communicate over vast distances. Whales, for example, use sounds or vocalisations to communicate their wants and needs to each other across the sea. Unfortunately, when it comes to understanding their high-frequency clicks and whistles, we have very little idea what they're saying.

But can we humans talk to the animals, grunt and squeak and squawk to the animals? The question was posed in the 19th century by another eccentric scientist with an interest in everything and anything: Francis Galton – a Victorian polymath who would possibly be more widely and fondly remembered today had he not also been something of an eye-watering racial supremacist who sort of invented eugenics. But for now, let's focus on dogs.

Galton had many odd interests. For example, he invented a boredom scale based on observing an audience's restlessness during lectures and public events. And during his own time as an undergraduate he came up with the Gumpion Reviver, a kind of portable dripping tap to be positioned over the heads of sleepy students to keep them alert. In the early 1880s he placed an ultrasonic whistle in the end of his walking stick and went around London Zoo, noting down which animals responded whenever he produced the sound. He reported that 'some curiosity is inevitably aroused by the unusual uproar my perambulations provoke in the canine community.'

Galton's experiments are the reason you can blow a dog whistle and your dog can look at you quizzically for a moment, then go back to intently sniffing the base of a tree. In *Game of Thrones*, the closest parallel to this would be the dragon horn, with which Euron Greyjoy returns to the Iron Islands, after many years away in the books.

Theon and Yara's marauding uncle claims he found the horn amongst the smoking ruins of Valyria, home of the dragonlords of old. It is bound with bands of red gold and Valyrian steel graven with enchantments, and has the power to 'control' dragons. Dany has had a certain amount of bother getting her dragons to behave nicely in public (and not devour other people's children), so possibly such a horn could be the answer to all her problems?

Essentially, the dragon horn is a fancy dragon-dog-whistle. And we all know how effective dog whistles can be. In a recent interview, Paul McCartney talked about adding a blast of one to the cacophony of sounds at the end of 'A Day In The Life', just to keep the pets of Beatles fans amused. Definitive proof that it's possible to use sound to devastating effect, and also to control animals (or at least get them to listen to The Beatles).

Against the Wall?

A group of men from all over the Known World are barracked together on the Wall. High- and low-born alike, they have sworn to defend the frontier of civilisation from the barbarous inhabitants beyond: those who will not 'bend the knee' to the rulers who govern absolutely everything to the South. Hang on, though, this isn't the great Wall in Westeros. It's Hadrian's Wall, built by the Romans in A.D. 122 to divide off the portion of Britain that they had managed to subdue and conquer – most of modern England and Wales – from the portion of it that violently resisted and defied Roman rule until the very last.

Hadrian's Wall runs for 80 miles coast-to-coast, across the very top edge of England: just south of the current Scottish Border. Constructed from squared stones, it stood 3 metres (9.8 feet) wide, and 5 to 6 metres (16 to 20 feet) high, and was built to last. Some of it is still standing now.

George RR Martin has said that Hadrian's Wall was the inspiration for his creation. He visited it with a friend, at the end of the day, when all the tourists were leaving, and felt the pull of staring out towards nothingness as those Romans must have done. (Well, he was staring towards Scotland, in fact, but I don't think he meant it rudely.)

When George RR Martin visited the Edinburgh Book Fair in 2015, Scotland was in the middle of the elections to decide whether to remain part of the UK or vote for independence. George was of course asked for his view and suggested a middle ground – whether the vote went to the Yes or No camp, Scotland should investigate constructing a massive wall of ice . . .

CHAPTER 7

MOTHER TONGUE OF DRAGONS

Where we learn a few real and made-up
languages, discover noun-verb constructs in
Dothraki, and eavesdrop on the talking trees.

~ Mother Tongue of Dragons: ~
The Magical Science of Inventing Languages

In 2012, 146 babies born in the US and 50 babies born in the UK were
named Khaleesi, to the horror of snobs everywhere. Of course, we know

this is a) excellent and b) in honour of Daenerys Targaryen, who is styled Khaleesi of Great Grass Sea, as well as Mother of Dragons, Queen of Meereen, and, in the TV show at least, Wearer of Wigs Extraordinaire. Naming your children after a popular fictional character is nothing new. In the early 20th century almost every Tom, Dick and Harry was named Wendy, after the popular heroine in J. M. Barrie's *Peter Pan*. However, this cherubic glut of Khaleesis is particularly interesting for two reasons. First, it isn't a first name at all. In fact, it's a title, roughly translating as 'Queen' or 'Wife of the Khal'. Second, the title is taken from an entirely fictional *Game of Thrones* language known as Dothraki.

Attitudes to made-up languages – like some of the languages them-selves – can be a bit odd. JRR Tolkien famously created not just indi-vidual languages, but languages that were interrelated the way that, say, English and French are. And everyone was mostly impressed (he was an Oxford University don after all, and some said that *Lord of the Rings* was really just a setting to show off the linguistic jewels he'd polished over the years). However, other forays into language creation have not garnered so much admiration.

One of the first full languages created for a fictional people was Klingon, for the Star Trek franchise. As languages go, Klingon has an interesting history. The first few words were uttered in *Star Trek: The Motion Picture*, and devised by actor James Doohan who plays Scotty (of 'Beam me up' notoriety). This sounds like a challenge for any actor, but during the D-Day landings in World War II, Doohan was shot six times on Juno beach, so some linguistic frippery was probably nothing he couldn't handle.

When Leonard Nimoy (Spock) was directing the third Star Trek film, *The Search for Spock*, he wanted the language to expand and not just sound like gibberish, so he hired linguist Marc Okrand to create a fully-fledged alien tongue, with a grammar based on those initial dozen or so words from Scotty. Thus a legend – or rather, a tantalisingly geeky opportunity to learn a proper alien language – was born. And the challenge was taken up with some enthusiasm by die-hard fans.

As any good nerd worth their eight-or-more-sided dice will tell you, Klingon has flourished over the years and there now exists a Klingon language version of Shakespeare's *Hamlet*. Nevertheless, 'Its vocabulary, heavily centered on *Star Trek*-Klingon concepts such as spacecraft or warfare, can sometimes make it cumbersome for everyday use,' notes Wikipedia – am I imagining a touch ruefully? (Though someone has requested a citation for this, perhaps highly controversial, suggestion.)

In her 2009 book, *In the Land of Invented Languages: Esperanto Rock Stars, Klingon Poets, Loglan Lovers, and the Mad Dreamers Who Tried to Build A Perfect Language*, Arika Okrent looks to internet message boards to detail the attitudes to those Klingonophones in the science fiction community. It makes for sobering reading. One particularly bad burn suggests that Klingon speakers 'provide excellent reasons for forced sterilisation. Then again, being able to speak Klingon pretty much does this without surgery.' Ouch. And this was on slashdot.org – the website billed as 'News for Nerds'.

In the interests of balancing this scurrilous libel, I refer you to the story of computational linguist Dr d'Armond Speers, who decided to bring up his baby son Alec speaking Klingon, while his wife spoke to their son in English (to ensure Alec would grow up bilingual/not too horrendously bullied). Dr Speers showed admirable commitment to the task in hand, even singing the Klingon Imperial Anthem, 'May the Empire Endure', at bedtime as a lullaby to his son (sample lyric: 'Our Empire is wonderful, and if anyone disagrees/we will crush them beneath our boots'). Little Alec picked it up. But unfortunately the limitations of the invented tongue gradually became apparent and the boy began to speak more and more English and less and less Klingon – though his dad observed that when he spoke the constructed language he did so perfectly, and never got confused with English words.

♫ DERRR
DUUURGH
DER DER DUHHH
DER DER DUHH ♫

Last time we checked in, Alec was a teenager and, according to Dr Speers, apparently retains little or no knowledge of, or interest in, speaking Klingon. Ah! Youth – and Klingon – are wasted on the young.

The Dothraki language had a similar genesis to Klingon. In the novel cycle, George RR Martin had already used a handful of Dothraki words, based loosely on ideas of how a mix of Arabic and Spanish might sound to someone who spoke neither tongue. So, when *Game of Thrones'* producers came to seek out new words, and boldly go-build-a-syntax where no-one but GRRM had gone before, they already had a few (*at* or *akat*) ideas in mind. The language had to fit in with phrases readers already knew, but it also had to be easily learnable and pronounceable for the actors (who actually rehearse their Dothraki lines in English, to get the sense across). After two months of competition between members of the Language Creation Society, two punishing rounds had whittled down the would-be Dothraki-creators to five Word Warriors. Of these final five, the triumphant Stallion that Mounts the World (or at least the Dictionary) was announced as David J Peterson.

Interestingly, the creator of Dothraki was inspired not by *Star Trek* but by watching *Star Wars*. He recalls a moment in his childhood, watching the scene in *Return of the Jedi* where Princess Leia, disguised as a bounty hunter, rescues Han Solo from Jabba the Hutt's palace.

Chewbacca is her pretend captive bounty, but to add to her exotic, foreign disguise, Leia pretends to negotiate a price for him in a language unknown to Jabba (and, apparently, to anyone who has given a thought to how human languages work).

Using the in-on-it C-3PO as Jabba's interpreter, Leia commences her brief but not uncomplicated arbitration using more or less the same word and subtle variations on it: 'yata'. This prompted the young linguist-to-be David Peterson to ponder in what language Leia could say the same thing twice, yet the second time it had a different meaning. This was the starting point for the fascination that would ultimately lead him to the world of *Game of Thrones*, and the utterances of the Horse Lords of the Great Grass Sea.

But maybe – maybe – this double/triple meaning was the mark of a truly alien language? (Or maybe – maybe – this just wasn't part of the *Star Wars* script that had received a great deal of attention, and Leia was therefore tasked with talking gibberish to humans and aliens . . .?)

Whatever the truth, Leia's alien bounty hunter routine reflects on an interesting and crucial problem relating to our understanding – or incomprehension – of a language. If you turn on your TV right now and hear a character on a show speak only a short line or two of strange dialogue, is there a way for you to tell what the language *is*? It could be a created language, like Klingon or Dothraki; it could be a natural language that exists and is spoken natively in our world that you don't know and haven't heard before; or it could be utter, utter gobbledygook. There really *isn't* any clear or easy or scientific way to tell.

We humans have evolved all kinds of sophisticated ways to use or misuse, illuminate and obfuscate in the tongue we have mastered, but we are completely clueless when faced with languages we don't know. Take, for example, the story of the silent illusionist Chung Ling Soo – 'the marvellous Chinese conjuror' – considered the world's greatest magician, at the beginning of the 20th century. He was a man who pioneered new ideas in magic, inspiring friends like Harry Houdini.

Soo's most notorious and dangerous illusion was the dramatic Bullet-catching trick. 'Defying the Bullets' involved catching two bullets in his teeth. Not surprisingly, given the risk involved, he performed it very rarely. The bullets were selected in full view of everyone, marked by the audience, then loaded into the muzzle of a gun and fired directly at the conjuror. In 1918, while performing 'Defying the Bullets' before an audience at the Wood Green Empire in north London, Soo spoke his first – and last – ever words on stage. The shots were fired as usual, but then, twisting his robed body awkwardly and sending the porcelain 'catch' plate smashing to the ground, the conjuror exclaimed, in perfect English, 'Oh My God. Something's happened. Lower the curtain.'

This exclamation from the previously silent performer was extraordinary for a number of reasons. The first was that Chung Ling Soo was actually William Robinson, an American of Scottish descent who neither spoke Chinese nor had any Chinese heritage. As Jim Steinmeyer describes in his biography of the magician *The Glorious Deception*, in order to maintain the illusion that came with his increasing success and fame, Soo developed a surreally complex method of hiding his identity behind unknown languages. He employed his Japanese friend Fukado 'Frank' Kametaro, who was a fluent English speaker, to act as his assistant and translator, when necessary for publicity reasons. When Robinson was required to 'speak' to the press, the reporters would ask questions in English, understood by both Kametaro and Soo. Kametaro would 'translate' the question into his improvised Chinese (he didn't speak it either) and then Soo would reply in his own, extemporised version of Chinese. Kametaro would acknowledge this and reply in English on Soo's behalf. So, two men who both spoke and understood English were creating two separately fake versions of a language neither knew in the slightest. You do have to wonder if they both always kept a straight face.

Extraordinarily, everyone was fooled. The illusion of fluency worked so well, of course, since the magician's company never appear to have encountered any Chinese-speaking reporters. A trick like that wouldn't work nowadays. Even in the pre-internet days, if you spotted a mistaken, phoney, or poorly-devised language you could only tell your friends about it – and they might or might not be interested. But nowadays, as soon as an episode of a TV show is broadcast, or a film is released, fans can rush online to admire and pick apart every nuance of the production – including the languages of the fictional world.

~ Constructed Languages ~ V Real-World Languages

Once upon a time some experts believed that as many as 12,000 languages were spoken around the world. Approximately 6,000 languages remain today, but by the end of this century, the consensus is that the number of languages in use will be much smaller than it is now. Some linguists reckon that in 100 years' time, 90% of the world's languages will be gone. And perhaps three dominant tongues – Mandarin, Chinese, English, Spanish – will rule the globe.

Meanwhile, in our mind-boggling era of globalisation and cultural exchange, constructed languages like Dothraki are created and learned within a few years, while real-world languages which evolved over millennia are dying out. And while many of us have names with meanings in languages ancient or modern that we don't speak ourselves, in future it may become commonplace to give children names that have meanings to select groups of people, or even just to the parents themselves, in a language they've lovingly created, along with their child.

Where could all this end? Could the number of created languages one day overtake the number of natural languages in our world? It's an intriguing possibility. (But, as the Dothraki don't say, it *isn't* known . . .) And what status do languages created for other worlds have in ours? In April 2016, the *Guardian* reported that a lawsuit currently before the US federal courts was dealing with this very question – disputing the use of the Klingon language in a new unauthorised crowd-funded Star Trek film. According to the *Guardian*, Paramount's attorney, David Grossman, argued it was 'absurd' to say that Klingon exists as an independent language. The Language Creation Society, on the other hand, filed a friend-of-the-court brief in support of the fan-made film, backing up their attorney's assertion that 'the [Klingon] language has taken on a life of its own'.

By contrast, the Dothraki language has a website providing a fairly comprehensive dictionary for fans, with over 3,000 words (many of which have never been heard in the show, and maybe never will be).

Thus Dothraki is gradually becoming a living, breathing language with a life of its own. It's even featured in the US version of *The Office*. And when the storyline on that show required some Dothraki words that hadn't featured in *Game of Thrones*, David Peterson was all too ready to help out. As the linguist explained on his blog, *The Office* involved one of the characters, Dwight, conjugating a noun–verb compound to do with throat-ripping. *Foth aggendak* – I throat-rip; *foth aggendi* – you throat-rip – you get the idea. Noun–verb compounds hadn't been called upon thus far in either George RR Martin's novels or the TV show, but hey, Peterson liked what the *The Office* had done in the way of noun incorporation. So he decided to canonise this linguistic innovation and henceforth this type of Dothraki noun–verb compound is known to language-learners in our world as a Schrutean compound, in honour of *The Office*'s Dwight Schrute. The mind- and world-bending possibilities of a constructed language!

Peterson has also taken a rather personal and romantic approach to the tongue of the war-like horse warriors. In any language he creates, he always includes his wife's name – Erin. For a less sweet-natured language creator, this raises the possibility for some interesting score-settling. But Peterson has '*Erin*' translated as 'good' or 'kind' in Dothraki. Aww! He also included his much-loved sadly deceased cat's name, '*Okeo*' to mean 'friend'. Pass me my '*arakh*' [fierce sword-y thing], I think there's something in my '*tih*' [eye] . . .

~ 'It's in the Trees, It's Coming!' ~

Back in the 1980s, Prince Charles, heir to the British throne, announced that he talked and listened to the plants and trees in his garden. Understandably, this delighted the world's press, and their happiness only grew, as in subsequent interviews Charles continued to fertilise the notion of this rather literal *Gardeners' Question Time*. 'I got a lot of flack for a lot of things. I mean potty this and potty that, loony this and loony that,' he told the BBC. Quite.

Another heir to a great kingdom, Bran Stark faces an arguably parallel dilemma as he too can commune with trees, and this is vital as he is a 'greenseer' – someone who can experience visions of the past and future of his world by connecting to a natural network of living plants.

Bran's guide and mentor, Bloodraven, helps him to master his supernatural powers. When we first meet Bloodraven he's seated on an eerie weirwood throne, a tangled nest of roots that almost seems at one with his withered limbs. In the world of *Game of Thrones*, anyone who actually has a throne commands our attention, yet Bloodraven's seat couldn't be further from the much-desired Iron Throne, made by the first Targaryen king out of the beaten and misshapen swords of his enemies, and smelted together with the fire of dragons. (It is extremely uncomfortable to sit on, by all accounts.)

Bloodraven sits dreaming upon his woven throne of weirwood, which has become a part of him, encircling him like a nest, embracing him like a mother. This natural and gentle nexus of power is the antithesis of the one above ground in Westoros. Yet what is Bloodraven's kingdom of trees actually like? Is it as weir[d] and spooky as we suppose? And in our world, what could a kingdom of trees actually achieve?

The old gods of Westeros and the Children of the Forest use a network of ancient trees to send and receive information across the Seven Kingdoms. Some of these trees literally have eyes; they have faces carved into them and are known as 'heart trees'. They seem generally benign, but from the start some readers and viewers have wondered about the motivations of the woods. Do they have Bran's best interests at heart?

A sentient tree manipulating a creature for its own ends is certainly something the acacia-tree dwelling ants of South America are familiar with. As the name suggests, these ants live and die entirely on the acacia tree and dutifully protect it from weeds and also attacks from other insects and even hungry goats. In return, the tree provides the ants with food and a place to live. So far, so wood. A classic 'symbiotic' relationship, biologists would say, that benefits both parties as they work together.

But recently, scientists studying the relationship discovered the acacia has a cunning plan to make sure this little arrangement never comes to an end – that the trees and the ants work, rest and play together *for ever and ever*. In order to make sure the ants continue to see the good for the trees, the acacia produces a highly addictive, sugary sap, which not only tastes good to the ants but also contains an enzyme that physically alters them so as to ensure they can never digest *any other* type of sugary sap from any of the other trees ever again. From their first hit of its sap, these ants are addicts who can never get clean, even if they want to. (We have the basis for a future dystopian novel about fast food mega corps right here . . .)

German plant-biologist Dr Martin Heil has studied the relationship between the ants and the trees. 'It was surprising to me that the immobile, "passive" plant can manipulate the seemingly much more active partner, the ant,' he says, in an article for *National Geographic*.

An eco-system that is self-aware? So far, so *Avatar*. But just as the weirwoods of Westeros communicate with each other and send information across many miles, so the trees in our world have fast and efficient ways of reaching out and 'talking' to each other. They just possibly don't discuss the politics of the Seven Kingdoms when they hook up.

A complex and extensive underground network of fungi connects trees and plants, enabling them to share information and help their neighbours, across many miles. Fungus expert Paul Stamets called these fungal networks 'Earth's natural internet' in a 2008 TED talk. Logging on to the 'wood wide web', there appears to be a constant 'back and forth' between the trees. They send each other nutrients such as carbon and nitrogen when needed. Old trees that seem to know they are dying will send packets of their own nutrients out to young trees that are just establishing themselves, to help them get started in life. The interconnectedness of forests and woodlands is such that some scientists think we shouldn't think of them as separate trees and plants at all, but as a single, sentient living organism.

OK, so trees help each other out with the basics of life, but weirwood trees that warn of dangers – that's still fantasy, right? Well, maybe not.

Scientists studying tomato plants have discovered that communications on the wood wide web can be more sophisticated than just an exchange of resources. These plants warn each other of incoming pests; when they have succumbed to pesky aphids they tell others to beware, and the plants receiving the message beef up their immune systems accordingly to avoid harm. Almost like those helpful friends who forward you an email saying 'Do not fall for this scam. He is not really a deposed prince trying to regain his country, but a guy who works in a mobile phone shop in Luton.'

We'll have to wait and see what happens with Bran and Bloodraven and indeed Prince Charles as they explore their unique relationships with plants. But can trees communicate valuable and complex information to each other? Yew better believe it . . .

CHAPTER 8

REAL MAGIC

Where we discover if magic exists and how to fake it; give the
gift of death; and make fantastical creatures in the lab.

⌒ Magic Returns to the World ⌒

At the beginning of our story, after a long absence, magic is returning
to the Seven Kingdoms. We see the signs everywhere – the mysterious
burning glass candles, the hatching dragons, the sawing women in
half. (Oh wait, actually no – that wasn't magic, that was Joffrey).

While we ourselves may still be waiting for such enchantment to return to our own humdrum world, there was a time in Western Europe, during the Middle Ages, when something very like actual magic began to zhuzh up people's lives.

Historian Professor Owen Davies tells me about the new 'magical' knowledge that began to change Europe during this period.

After long years of warfare, the Black Death pandemic and raids by everyone from Vikings to Visigoths, many of the ideas, technologies and literary works that had stirred civilisation during the Roman period were lost in Europe. The sophistication of the Romans had built on that of the Greeks before them in the fields of geometry, astronomy, architecture, philosophy and more, but now in Europe those ideas were in ruins. The books that contained them had been lost or destroyed, bar a few pages copied out that turned up here and there, and even the memory of the knowledge was half-forgotten.

But then, new people and new ideas began to arrive. In 1453, Constantinopole (now Istanbul) was the last bastion of the Roman Empire, but under the command of 21-year-old Sultan Mehmed II (known as Mehmed the Conqueror) it fell to the Turkish Ottoman forces. Scholars and philosophers who had lived in the city fled – many to Italy. And when they arrived in Italy they brought with them many texts and works that hadn't been seen in Western Europe for hundreds of years.

Ancient texts treasured and pored over were available again, for those who dreamed of discovering the fountain of eternal youth, and how to turn base metals into gold, and how to cast spells using Babylonian Love Magic.

The books being rediscovered contained useful facts but they also contained wonderful fictions. What was and what wasn't true? So much seemed possible!

In southern Spain, the Moorish civilisation flourished. And Muslim scholars – who had been the main custodians of Greek philosophy – were another conduit through which more knowledge flowed back into Western Europe. Knowledge of how to do anything and everything, from organising libraries to cultivating fruit trees, reappeared in the world. And the effect was extraordinary. Imagine travelling back in time to someone doing their homework in the early 1990s and showing them today's Wikipedia.

Jewish, Islamic and Christian thinkers influenced and were influenced by each other during this period, which seems to have enriched each of the religious traditions. But in addition to religiously sanctioned ideas, the classical works that arrived back in Europe also included works on magic itself. The Greeks had been very interested in this, taking ideas from the Babylonians. Religion and magic, magic and science began to cross-pollinate . . .

The Magical Power of a King's Flesh and Blood

———·———

Melisandre is deeply into blood magic. According to ancient lore, blood forms the basis of powerful potions, particularly when it belongs to royalty, which is why she gets a glint in her eye whenever a member of the royal family gets so much as a paper cut. Melisandre's magic seems both terrifying and impressive. At one point we see her gorging three leeches with the blood of King Robert's (bastard) son – the parasites are rather unflattering stand-ins for the three 'false' kings, Rob Stark, Balon Greyjoy and Joffrey Baratheon. Melisandre tosses the hapless leeches into the flames and before long the royal rivals are all dead. Coincidence? Hmmm.

The Red Woman's love of noble blood is not without historical precedent in our world. In medieval Europe, for instance, it was believed that intimate contact with royal flesh could achieve wonders – but without the royal in question dying a prolonged death. In England and France during the Middle Ages, many people thought that a royal personage could channel divine powers to miraculously cure grim diseases. English and French monarchs were sought after to cure a skin disease known as 'scrofula' or 'the king's evil'. And throughout Central Europe in the 16th and 17th centuries, a belief that the ruling House of Habsburg dynasty could apparently cure a stammer with a full mouth kiss thrived (although having seen portraits of the Habsburg family, you might be better off sticking with the stammer).

Belief in such regal cures probably persisted because these diseases rarely killed people, and sometimes improved of their own accord – thus giving credence to the healing power of a royal grope. Monarchs who were a little less keen on getting their hands dirty would finger, bless, and then distribute cramp rings (known as 'touch pieces') to the sick and needy. (Often these rings were made of solid gold, so worth having either way.)

In the 18th century, the anti-touchy-feely monarch George II firmly put an end to getting hands-on with his sickly subjects. Not much fancying the idea of his own 'magic' flesh and blood, he instead favoured and promoted a different approach – inoculation. A forerunner of modern-day immunisation, the process involved trying to prevent diseases such as smallpox by deliberately injecting smallpox pustules into someone to produce a less severe infection that then induced immunity to a much more serious dose.

∽ Divided by Magic ∼

Westeros is divided in many ways, but one of the more subtle and inter-esting divisions relates to magic. Some of the characters have long been practising it whilst others fear and wish to destroy it. Some scoff at magic, believing it to be nonsense, others are convinced that it is living, burning and freezing in the realms of ice and fire. And as readers and viewers, we tend to agree with them. What we see feels like real magic.

In one 'magic' corner, we have dragons and White Walkers and (presumably) all the fantastical beasts, like 'grumpkins and snarks', that Tyrion describes. In the opposite 'unmagic' corner, there are the Maesters, those scientists, historians and postmasters who advise the Westerosi nobility, and hate dragons like there's no tomorrow (which of course there probably won't be, if the White Walkers get their way).

Of course, the magic in *Game of Thrones* is not about pulling rabbits out of top hats, making coins appear behind ears or guessing the right card. Instead, it is unpredictable and those who use it are not necessar-ily shielded from its consequences. For instance, Jon Snow asks the wildlings, if they have a magic horn that can bring down the wall, why don't they just use it? The wildlings' reply is as instructive as it is chill-ing: there's no easy safe way to use magic. They say it's like having a sword with no hilt – how can it be safely grasped?

Not everyone is a fan of magic. In the Seven Kingdoms the stakes are too high, with one person's life saved at the cost of another person's death.

Lord Varys the eunuch, for example, clearly hates magic. He tells Tyrion Lannister the horrific story of how as a young boy he was castrated by a sorcerer and watched his genitals burn in a blood magic ritual. Having served his purpose, Varys was then thrown out into the street to bleed to death.

Still angry, still troubled, Varys reveals that he still dreams of that night and the voice that came from the flames as his flesh burned. What was it? A god, a demon or simply a ventriloquist? While the eunuch recounts his tale, he unwraps a large box. And we discover that, after all these years, he has tracked down the responsible magician who is now powerlessly *trapped inside the box*, like a spider under a glass. It's a dramatic reveal, darkly reminiscent of David Blaine's London perspex-crate-based stunt beside the Thames, but with a smaller if equally belligerent audience.

But Varys isn't the only one you'd hesitate to bring along to watch David Copperfield. The Maesters are also rather down on supernatural happenings. On the night before they officially become Maesters, each of them must complete their training by spending a night trying to use various spells to light a long stick of obsidian referred to as a 'glass candle'. Such candles are kept as novelties and associated with quaint stories about how they once burned with a strange, colourful light that enabled sorcerers to communicate half a world away. The trainee Maester's tireless, but ultimately unsuccessful efforts are designed to demonstrate that magic is not real, that there are limits to what it is possible to know.

However, despite the deep-seated unease and scepticism, there is no stopping the rise of magic in the Seven Kingdoms. The glass candles are beginning to light up again, of their own accord, as magic slowly returns to the Known World.

∽ Giving the Gift of Death ∽
(Keep the Receipt)

Many significant occasions in life are marked by an exchange of gifts, in both our world and the world of *Game of Thrones*. But it is often hard to find the appropriate gift for any occasion – marriage, festival, or the opening night of a friend's play. The Faceless Men of the great city of Braavos have solved this dilemma. Not for them all of that worrying in Marks & Spencer over which artisan home ice-cream maker looks the best and most expensive the afternoon before a big event. Nope! The Faceless Men always give the same gift, and it works for every occasion: the Gift of the Many Faced God. If that sounds spiritual-yet-practical, remember this is *Game of Thrones* – the Gift in question is . . . Death. (Durr – obviously . . .)

The Faceless Men – more correctly The Faceless Persons, as women can totally join – are a sort of assassin's guild. They are hugely skilled at what they do, and can be hired at great cost to off a well-guarded enemy, or bring a swift, merciful end to the suffering of a dying loved-one.

But death is not a gift they go around giving to just anyone. Oh no! The Faceless Men get very angry indeed with Arya, who fights in anger and recites her kill list each night. In the books, The Faceless Men have in-depth discussions about how they cannot kill those they know. This attempt to rationalise killing, to rob it of all emotion, can appear odd

205

to those of us who live in countries that don't have the death penalty, or who personally oppose it. But it also seems to be a very human thing to do.

Arya sees the assassin's company of Faceless Men sitting around the dinner table, planning who will kill who. Y'know, as you do. Some of them cannot kill the men or women they've been asked to kill because they recognise their names, and know them. Others agree to kill them instead because they don't know them.

These scruples about killing someone you know extends to assassins in the real world as well. Take, for example, criminologist Saul Alinsky's dissertation on Al Capone and his associates, in the 1930s. Alinsky spent time with the legendary gangster and his associates, and got a unique and detailed insight into their murderously criminal activities.

In an interview for *Playboy*, he remembered looking over the mobster's records and spotting a large payment to an out-of-town killer. (Even gangsters can't escape admin. Who knew?) When Alinsky asked why it was necessary to bring someone in to kill, when there were 20 killers on the payroll, he was told that this was because some of the targets may be known to them – apparently, the mobster contact seemed genuinely shocked at his callousness in asking such a dumb question.

Arya Stark runs away to join the Faceless Men to learn their assassinatin' ways, fuelled by anger against those who have destroyed her family. Entirely reasonably, she wants to kill every single cruel and murderous individual who has messed with her family, and we cheer her on in this. What she isn't expecting is to have to practise a sort of kindly euthanasia for those who are suffering.

A father brings a young girl to the House of Black and White so she can find release from her sickness which cannot be cured. Never a show to shy away from showing grim stuff happening to small children (Shireen!), we see Arya give the girl, not much younger than herself, a drink from the fountain in the centre of the temple of the Many Faced God. And the girl duly dies. This would be traumatic for some, but Arya, who, as we know, is probably an unfeeling psychopath by this stage, seems fine with it. And we love her anyway.

It is not clear whether the fountain is supernatural. But the powders and potions that Arya learns to mix in Braavos, whilst blind, must be learned by smell and touch, implying they perhaps contain 'real' chemicals.

In our world, the medicalisation of executions – and the production of a painless liquid poison – has been a controversial one. The drug of choice for death by lethal injection in the United States – sodium thiopental – was invented in 1934. It was controversial from the start. Used as an anaesthetic to treat the wounded after the Pearl Harbor attacks in the US, it was implicated in the deaths of some of those who were operated on – though subsequent research suggests the drug probably wasn't to blame for the deaths. In small doses, it was used as the 'truth serum' you hear about in Hollywood movies of the 1940s and '50s. All but one of the 35 US states that carry out executions used the drug in lethal injections. But in 2011 the supply dried up, as the sole American manufacturer of the drug announced it was ending production. And manufacturers in Europe already faced a ban on exporting the drug to the US, as capital punishment is not permitted within the European Union.

Faced with such obstacles, US death row states started using an untested drug, pentobarbital, in executions (the drug is licensed to treat the most stubborn forms of epilepsy). But in May 2016 the drug company Pfizer imposed sweeping restrictions on the distribution of all its products, to ensure that they couldn't be used in lethal injections. And thus 'all FDA approved manufacturers of any potential execution drug have now blocked their sale for this purpose' according to a representative of the Human Rights Charity Reprieve (as reported in the *New York Times*).

The strain on the death row supply chain spells a decline in judicial killings in the US. According to the Death Penalty Information Center (it's a thing), the scarcity of injection drugs has contributed to the number of scheduled executions dropping dramatically – 98 were carried out in the US in 1999, as opposed to 28 in 2015.

The Ultimate Ride

While death penalty states in the US contemplate bringing back firing squads and electric chairs, in the absence of available medical drugs, sodium thiopental and pentobarbital are still used to end life in European countries where euthanasia and assisted suicide are legal (or at least tacitly permitted). But a 'humane' voluntary death needn't be boring.

For those whose answer to Syrio Forel's question 'What do we say to the god of death?' is, in fact, 'Today', Julijonas Urbonas, a PhD student at the Royal College of Art in London, has designed the Euthanasia Coaster. N.B. If you are reading this in a theme park you might want to skip ahead a chapter right now.

A hypothetical death machine, Julijonas' rollercoaster is, he says, designed to kill passengers 'with elegance and euphoria' and was inspired by the president of one of the world's oldest rollercoaster manufacturers, who described the ultimate coaster as leaving its passengers dead at the end of the ride.

After travelling up a steep incline (so at least you get a good view as you go out) passengers shoot down the 510 metre slope of the rollercoaster's hill, at 100 metres per second (g-force 10). At the bottom of the slope they quickly go through seven inversions that would inflict the equivalent of 10 times the force of gravity on the world-weary riders, causing death through cerebral hypoxia (a lack of oxygen to the brain).

～ Snarks and Grumpkins ～ and Fairies — Why We Want to Believe

When Jon Snow speaks of the role of the Night's Watch in protecting the 'civilised' world, Tyrion Lannister interrupts him mockingly, suggesting that the only things out there, in the frozen North, are grumpkins and snarks. The White Walkers are simply another mythical monster that wet nurses frighten children with. The men who defend the Wall may think that all Seven Kingdoms are counting on them, but no sophisticated Westerosi believes in fairy tales.

Almost from the start, readers and viewers know that Tyrion is mistaken: the magic and myth surrounding the White Walkers *is* real: the 'Others' are a greater threat than the War of the Five Kings. But the entertainment comes from watching belief and scepticism shape the world of *Game of Thrones*.

We know very little about Tyrion's snarks and grumpkins, but they seem to be a little like the fairies of our world, even down to grumpkins being able to grant three wishes. The mysterious Children of the Forest, the first inhabitants of Westeros, are close to the 'fairy' tradition in our world too. And it is they who created the White Walkers with magic that then went bad. George RR Martin compares the 'Others' to the Aos Sí, a supernatural race from Irish and Scottish mythology who live in ancestral burial mounds.

In our world there are many who believe in the existence of fairies, sprites, elves and other hidden people. Fairy tales were once something for everyone. The Grimm brothers collected stories to entertain adults. And storytellers around the world passed the long nights with their narratives of 'Once upon a time . . .' But throughout the 20th century, believers and sceptics argued over the effect of a belief in the supernatural.

The most famous case dividing the two sides is the Cottingley Fairies, a series of five photographs, apparently depicting two cousins interacting with fairies.

In 1917 the village of Cottingley in Yorkshire was home to two young girls, Elsie Wright (aged 16) and her cousin Frances Griffith (aged 9). The cousins would spend hours together playing by Cottingley stream and would come home to tell Elsie's parents of the fairies they encountered there. Not being taken very seriously by their family, one day Elsie borrowed her father's camera and later triumphantly returned having captured 'proof' of the fairies. Elsie and her keen amateur photographer dad developed the pictures from glass plate negatives in his darkroom. Elsie's father still wasn't convinced, but her mother was less doubtful and took the pictures along to a lecture she attended that touched on the subject of fairies. When experts and learned men saw the pictures, the story really began to take off, with the master of detective fiction, Sir Arthur Conan Doyle, writing about the fairies in a popular magazine's Christmas issue.

Because of the nature of the early plates used to photograph the fairies, it wasn't possible at the time to assess the images in great detail, in the way that photographs of apparently strange phenomena are routinely examined today. Nowadays we can 'blow up' images to study every minuscule portion of them, but 100 years ago, when confronted with fairy folk, it wasn't as simple as just taking a really close look at the pixi[l]ies.

At Conan Doyle's prompting, various photographic experts in the 1920s (including some from Kodak) vouched for the photos. They couldn't go so far as to say they were genuinely capturing the images of fairies, but they did not see any obvious signs of being 'faked' (though they didn't go so far as to say 'it's *definitely true* – fairies are real, everyone!'). Others poured scorn on the very idea that anyone could possibly believe in fairies at all, with or without photographic proof, so the controversy continued.

Doyle was famous for writing the deductive reasoning of Sherlock Holmes into existence, but his keen interest in the supernatural lost him good friends. Harry Houdini had previously fallen out with Doyle over his fervent belief in séances. Still, Doyle's belief that the photographs were authentic was shared by many.

The photographs appeared at a time of great upheaval in Western Europe. They were taken in 1917 while World War I was being fought, and captured the public's imagination when they were published in 1920, two years after the war ended. The 1920s saw a collective yearning for the lost sons and fathers and brothers and husbands who had violently perished in the fighting. When the girls' photographs appeared, they offered up dancing fairies gathering flowers to a world that longed for the innocence of a numinous experience after so much bloodshed and horror.

But the end of the story is in its beginning. When first shown the photographs, Frances's father teased her that she was messing about photographing 'bits of paper'. American magician and arch-debunker of the supernatural James Randi returned to that idea in the late 1970s, identifying and widely publicising the similarity between the 1917 fairy photographs and some illustrations of dancing, elegantly draped figures that appeared in *Princess Mary's Gift Book*, published a couple of years before the fairies first materialised. Here are the same dresses, the same poses – unmistakably copied from the book. It's clear we are looking at little cut-out drawings.

As first reported in the *New Scientist* magazine in 1978, Randi believed he had detected 'strings' holding the paper fairies up by enhancing the images 'using the latest technology'. However, these strings are probably no more real than the fairies themselves. As Elsie eventually admitted, the girls had cannily encouraged the well-drawn little folk to remain upright and spritely using *hatpins*. You can spot the point of one of said pins in the photo of Elsie and a small gnome. Conan Doyle noticed this 'dot' and, from its position on the gnome's midsection, assumed it to be a belly button, thus indicating to his mind that fairies give birth to babies with umbilical cords, just like humans do. (Had the tip of that pin ended up elsewhere, who knows what other conclusions the creator of Sherlock Holmes would have drawn about the fairy anatomy?) But it seems we all sometimes see what we want to see – sceptic or believer.

Both women hinted at different times that they had seen something, or somehow photographed their 'thoughts', before Elsie finally admitted that the photos were a hoax. Frances, on the other hand, maintained that the fifth and final photo they took was genuine. It shows the fairies alone together in the grass; they appear to be semi-transparent, most likely caused by a double exposure.

Prompted by the photo that shows one of the fairies atop a mushroom, psychologist Professor Richard Wiseman discovered a botanical map of the Cottingley area that dates from Elsie and Frances's time, showing the distribution of mushrooms in the area round the stream. Many of them were magic, the type of mushroom containing the psychedelic compounds psilocybin and psilocin, which famously cause hallucinations.

Given that for obvious reasons the two women were reluctant to expose their hoax and each other, it's unlikely that they would have made things even more weird by admitting to being off their faces on hallucinogens as children, and it's also unlikely they would have been deliberately eating the mushrooms for this reason. But it does *possibly* offer an explanation of why both Elsie and Frances maintained until the last that even though the photos were faked, they really did see fairies down by the stream.

♫ DERRR
DUUURGH
DER DER DUHHH
DER DER DUHH ♫

There's a longstanding link in European folklore between fairies and mushrooms. Fairy rings – circles ranging in diameter from less than a metre to over 10m (33ft) – are still to be seen on open grassland and in woods. A fairy ring, also sometimes referred to as an 'elf circle' (but never as a 'pixie hole') is certainly a charming sight. Traditionally these circles are said to have been made by the little folk dancing in the round, but we now know in fact they're caused by the presence of the fungus mycelium in or underneath the outlined arc. Around 60 different species of mushroom can grow in this ring pattern.

We see an echo of this link between magical folk and consuming hallucinogenic substances in the world of *Game of Thrones*, with the visions of Bran Stark. When Bran eventually finds his way to the caves where he meets his spiritual guide and teacher Bloodraven, as well as the fairy-like Children of the Forest, he is given a bowl of weirwood paste, made from the sap of the ancient trees, which when he eats it gives him visions of the past and future awakening his belief in his own abilities as a 'greenseer'.

Whether or not Elsie and Frances were tripping on 'shrooms (let's not judge) there's something strange and beautiful, moving even, about the story of these clever mischievous young women who brought some magic into a world that so desperately wanted to believe in it.

In an interview in 1985, a few years before her death, the elderly Frances gave an interview to Arthur C. Clarke for his popular *World of Strange Powers* TV show. She explained that as a child she felt too embarrassed to tell the truth after the stories and photographs of fairies she and her cousin created captivated a 'brilliant' man like Conan Doyle. 'I never even thought of it as being a fraud,' she said. 'It was just Elsie and I having a bit of fun and I can't understand to this day why they were taken in. They *wanted* to be taken in.'

The spectacle of people thirsting for spiritual fulfilment and purity after the turmoil of warfare and violent loss is a cycle that repeats itself, time and time again. We see it happen in Westeros with the rise of the Faith Militant, the 'extremist' arm of the Faith of the Seven, the major Westerosi religion. The revival of spiritual fervour is a direct response to the terrible and destructive War of the Five Kings. Even in times of peace, environmental psychologists have found that a belief in the spiritual may prove advantageous. Research shows that, whilst secular folk live purposeful meaningful lives, those with religion or spirituality tend to feel their lives restored and protected through difficulties, by believing in supernatural forces beyond their control.

∽ What makes a believer? ∾ Or, the Sceptical Smiths

We know magic is a real force in the Seven Kingdoms, yet for all the characters who believe in the Red Witch Melisandre's predictions, or the White Walkers, or the Children of the Forest, there are individuals in the world of Ice and Fire who are decidedly sceptical.

In our world, neuroscientists and psychologists have shown an abiding interest in studying why some of us believe in the supernatural while others scoff at it.

All kinds of intriguing theories as to why we do not accept an idea while others have been advanced.

It has long been known that believers in matters paranormal are prone to seeing patterns where none *necessarily* exist. Believers are especially likely, for instance, to see a non-existent ghost popping up behind Grandma Jane in their holiday photos, or to conclude that a dream bears an uncanny resemblance to the following day's news headlines and is therefore 'a sign'. Some scientists have taken a decidedly brain-based approach to explain this phenomenon, suggesting that this overly-enthusiastic approach to spotting patterns is due to having lots of dopamine (a neurotransmitter that carries signals between nerve cells) in the brain.

Support for this theory comes from a set of studies in which sceptical participants were given a drug that increased their dopamine levels, and then promptly started seeing coincidences and non-existent patterns during several tests. Prior to the dopamine-boosting shot the sceptics missed spotting some of the words and faces the researchers briefly flashed up in front of them on a screen. After they'd received the shot, they became more likely to notice coincidences and patterns emerging, identifying even the scrambled faces and words as the real thing. Given that the level of dopamine swishing around in your brain is determined, in part, during your time in the womb, this theory suggests that some people are indeed born to believe.

To me, one of the most interesting theories that came out of the studies is that those who are likely to believe in supernatural events are less likely to be familiar with probability theory. You can see how this might work. Let's imagine that you dream about winning some money and then wake up to a telephone call from your friend announcing that they have just hit the jackpot on the National Lottery. In reality, you dream almost every night of your life and so, just by chance, at some point some of these dreams will show some similarity to subsequent events. However, if you haven't encountered this aspect of probability theory then you might be tempted to conclude that your dream *predicted* your friend's lottery win. Of course, other aspects of psychology suggest whether you will conclude your claim to prophecy, with a reminder 'about that money I lent you in 1994'. But I digress.

The 'believers don't really get probability' is a nice theory, but is it true?

Let's consider a test study.

Imagine you meet Mrs Smith in the street and you know she's a mother of two (How? Perhaps she's wearing a T-shirt saying 'Mum of Two of the Year'). Anyway, then she introduces you to the youngster at her side. Let's call him Shakespeare Smith.

'This charming man is my son!' she says, beaming proudly.

What are the chances her other child is a girl?' When I first heard this problem I thought – easy – 1/2! Obvious! But I was wrong. There are four possible scenarios re Mrs Smith's two children.

1 Her first child is a girl and her second child is a girl.
2 Her first child is a girl and her second child is a boy.
3 Her first child is a boy and her second child is a girl.
4 Her first child is a boy and her second child is a boy.

We know that Mrs Smith has a son as he's here in front of us, wearing his 'Mrs Smith's Son' T-shirt (they're a remarkably literal family, the Smiths), so 1 can't be true. That leaves us with 2, 3 or 4 as possibilities. As two of these three possibilities give us 'the other Smith child is a girl' (let's call her Sheila Smith) the chances are 2/3.

Immediate, real world applications for this could, for instance, be the next time you meet a very attractive but attached person, you can do a quick bit of working out the chances on the back of an envelope before asking, 'Do you have a brother?'

This nice illustration of how probability theory shows how our intuition doesn't always match up against the maths. It is known as The Boy or Girl Paradox and was popularised in the 1950s by Martin Gardner, the much-loved mathematician, magician and Alice-in-Wonderland-aficionado.

These types of questions have been presented to believers and sceptics over the years and the results show that the 'disbelievers' tend to outperform the 'believers' time and again on tasks requiring statistical thinking or involving probability judgements. If you've never encountered probability theory you're more likely to believe in a supernatural explanation, as we humans just like things to be explained *somehow*.

Presumably the reverse could also be true. If you're brought up in a very rational environment, versed in statistics and numbers and measurements, and never really have much to do with thinking about the paranormal, you're more likely to find rational explanations for why you're having what seems like a particularly unlucky day. Let's imagine you wake up, it's raining, then on your way to work you're splashed by a bus, then the heel falls off your shoe, and when you get to work your boss tells you you're being made redundant. For a rational type, your reaction may well be to recall hearing that the municipal authorities in your area have recently cut back on spending on road maintenance, leading to more pond-like potholes when it rains (which it does a lot at this time of year), then ponder that your shoes were £4.99 in the sale – and maybe there was a reason for that? – before finally looking up the employment statistics for your job and realising the number of people earning a living in your profession has shrunk by 20% in the last decade. (You would probably never say to yourself *'I bet a witch is behind all this'*.)

Anyway if you got the answer right, congratulations. Take a bow. You are the rational type. If, like me, you managed to get the wrong answer then you are more intuitive, and therefore possibly more likely to believe in the paranormal. Saying that, if you do happen to dream about next week's winning lottery numbers, drop me a line?

∽ Gene Genie ... ⌒

If, *unmagically*, we could get hold of some fresh, fertilised, fantastic dragon eggs in our world, would it really be possible to hatch them with science? In 2006, the *Economist* ran a story about Dr Paolo Fril, chief scientist at the Gene Duplication Corporation, who was running computer models with the expectation of one day creating a whole host of mythological beasts, including dragons, gryphons and unicorns. As you might expect, certain aspects of this process were proving tricky. But, as the *Economist* reported, 'If he can get the dragons' respiration correct, he thinks they will set the world on fire.'

Many researchers thought the idea was a touch unfeasible, but many more noted that the 1st April publication date was maybe a clue about the true nature of the article. However, fast forward a few years to the present day, and a new gene-editing technology called CRISPR-Cas9 has come along. And suddenly the world is full of intriguing possibilities.

CRISPR-Cas9 can be described as DNA scissors. It's a gene-editing tool that allows scientists to snip out bits of the genome with molecular precision. These 'holes' can then be patched with another piece of genetic material, repairing or replacing what has just been cut out. In short, scientists now possess minuscule sewing kits that might, in skilled hands, one day be capable of tailoring the basic stuff of life into ever more fantastic designs.

While we wait for Professors Emmanuelle Charpentier and Jennifer Doudna to be awarded the Nobel Prize for Chemistry for developing CRISPR, today's Biology undergraduates at universities around the world are already using it for 'simple' genome altering projects. And this is just the beginning. It's hoped that the ability to 'edit' genes could be used for tremendous good in our world. For instance, we could alter the genes of the infected mosquito that carries malaria in its bite, rendering it harmless and thus wiping out a disease that kills around a million people each year (most of them young children). Viruses have a genome, and therefore in future we could perhaps use CRISPR to mutate HIV. A trial using patients' own immune cells adapted with CRISPR to target and destroy their cancers has recently begun.

And once the gene-editing genie is out of the bottle, we can start having some serious fun. In 2015, two bioethicists, R Alta Charo of University of Wisconsin School of Medicine and Public Health and Henry T Greely, of the Stanford School of Medicine, published an essay about the possible role of CRISPR in making science fiction science fact.

They ask, 'If you could wish into existence any animal, plant or microbe you wanted, what would you make?' (I went for 'a flock of tame golden hummingbirds', then 'Channing Tatum'.)

The bioethicists sensibly caution that CRISPR 'will not repeal the laws of physics – no flying horses – and biology may just not permit some variations – perhaps no large animals with functioning wheels'. (Boring!) But given continued advances in genome editing technologies and our understanding of how DNA works, pretty much any animal that could be created, could, well, be created. Which means it's only a matter of time before a billionaire decides to give his daughter a unicorn for her birthday, or someone has a go at designing . . . wait for it . . . a real dragon!

Exploring the premise further, the authors note that physics will almost certainly combine with biological constraints to prevent the creation of flying dragons or fire-breathing dragons (see Chapter One guys!), but a very large reptile that looks at least somewhat like the

European or Asian dragon – perhaps with flappable if not flyable wings? – that could be possible some day they believe.

Because of the legal complexities involved in regulating the brave new frontier of genetic manipulation (the issues are considerable, complex, and likely to be thrashed out in laboratories and courts around the world for years to come), work in this area may not be outlawed for quite some time. Which means that several Do-it-Yourself 'biohackers' are on the case, say Charo and Greely, 'exploring possibilities such as changing blue flowers back to their original white, or producing the protein needed for vegan cheese, all in their own homes or informal community labs'. Tasty vegan cheese today, dragons tomorrow? Who knows?

George RR Martin's fiction reveals his extensive interest in and knowledge of genetics. And *Game of Thrones* fans have learned to be a patient bunch. So, if the great man is taking a little while longer than expected finishing off the books, could it be that he's busy with some DNA, and some very, very tiny scissors . . .?

Geteticist's Question Time

———————•———————

Working day and night to create a fantasy-made-reality creature sounds great to me. But in order to design an animal, we have to know the principles involved, and at the moment our knowledge is pretty poor.

It's fair to say scientists who work with DNA and genetics are working in a paradoxical and strange world. Nothing is quite as it first seems. For instance – logically the more complex a living thing is, the more genetic material we'd expect it to have, right? Well – that's not really how it works. A simple raspberry has around 8% as much DNA as you or I. Fair enough. But a humble onion has more, much more, DNA than the Onion Knight Ser Davos Seaworth, or any other human being you care to mention.

Furthermore the way genes are arranged is not how you or I might expect, or how scientists studying the manipulation of genes might hope. To use an everyday analogy, when we look at the human genome a layperson would reasonably imagine that different genes would be grouped together in appropriately organised sections, like the aisles of a supermarket. All the genes that deal with eyes – colour, etc., – would be in the 'eye section' of the genome just like, say, all the cheese and cheese-related foodstuffs are in the chilled section of the supermarket labelled 'cheeses'. However, if you've ever worked in or visited a busy supermarket then it's easier to imagine the arrangement of the genome as resembling the aisles of extraordinary chaos at the end of a very long and very hectic Saturday night. Pick up a loaf of bread in the bakery and someone has shoved novelty *Top Gear* socks under it. The shelf where you expected to find air fresheners is empty, save for a lone multi pack of condoms, and some Stilton. A customer clearly in a hurry has left a toy telephone in the meat aisle, nestled among the packets of wafer-thin honey-smoked ham. How did *that* get *there*? Basically stuff is (literally) all over the shop.

Yet, *somehow* it all seems to hang together. Evolution has found a way. So this gene here may have a surprising way of combining with another, totally different gene there that seems to control something else entirely, but, somehow, through millions and millions of years of natural selection, it all works. Rather like if you just pop out on a Sunday morning, bleary-eyed in your pyjamas for a coffee, and stop by the supermarket to pick up a newspaper, but hang on! There's a carton of still-cold semi-skimmed milk left tucked inside your paper – and you'd forgotten you were all out of milk for your breakfast cereal. So you certainly weren't expecting it, but now you've got it, that's worked out well.

And hence the key issue with all this potential biological engineering is knowing where to snip and what to change when faced with a far from obvious or logical layout of possibilities in a genome. At the moment we're still at the stage of working out what everything does, rather than being skilful enough to do 'proper re-engineering'.

Geneticist Dr Jonathan Pettitt from the University of Aberdeen explains things further using car analogies, which gets enjoyably *Wacky Races*. So, start your engines! Essentially, he says, geneticists do the biological equivalent of removing a bolt or a wire in a car engine, and then see what happens to the ability to drive the car. When they remove a gene, nothing obvious seems to happen sometimes ('actually a lot of times,' says Dr Pettitt). But maybe that means 'we've just disabled the airbag, so won't see anything obvious unless we crash the car, or perhaps the heated seats stop working, and these make only a fractional difference to the experience of driving the car. Other times something catastrophic happens – we can work out what has gone wrong. The car won't go round the corner of the canyon road, but continues in a straight line right over the edge, before dropping like a stone'. There are, Pettitt explains, other cases where catastrophic things happen, but we have no idea why. It's the equivalent of you turning the car key and the engine instantly bursting into flames.

That's a problem.

So just because geneticists have molecular scissors to make whatever changes they want to, it doesn't mean they can just go in and make the alterations needed to produce a tailor-made genome. The University of Aberdeen's gene supremo Dr Pettitt confesses that 'we're in a worse state of knowledge than even the most clueless participant on *Project Runway*. Even something simple like engineering a specific eye, or hair colour, runs up against the problem that these traits are not determined by single genes, and the interactions between the gene variants involved can give unexpected results.'

Surely, though, geneticists are already mastering the big, obvious stuff, like understanding and altering our height and weight? Nope, says Pettitt. It seems that here the issue gets even more problematic. As we understand them, both height and weight have strong genetic components. Eighty per cent of the variation in height, for instance, is down to genetic variation, with the remainder being composed of other variable factors, nebulously described as 'environment'. So in theory it might be possible to use gene editing to engineer a baby boy or girl that would grow up to be as tall and mighty as, say, Gregor Clegane, aka The Mountain That Rides, or Brienne of Tarth. But it's complicated, because there are over 400 genes that contribute to the variation in height in a population, and each one makes a tiny contribution.

Dr Pettitt again: 'It's true that there are some genes that we know a lot about, the ones that control development, and they're just the ones we might want to change to create a dragon. We still don't know enough yet to do this with any precision, but, even worse, genes interact with each other in ways that are difficult to predict. To use the earlier *Project Runway* analogy, it's as if you changed the shape of the collar of a dress only to now find that inexplicably all the buttonholes have disappeared.'

But perhaps an entirely different, more direct approach may yield results. In 2016 an invitation-only meeting was held at Harvard Medical School. On the agenda, The Human Genome Project-Write. The aim of HGP-Write is breathtakingly ambitious – to construct whole genomes, not just for humans but for other animals as well, and then to get this synthetic DNA working in living cells. So, as Dr Pettitt points out to me, that gives us another, daring, if for now distant, option – to design the entire genome of our dragon *from scratch*.

Why might this be a better option than gene editing with CRISPR? Well, if scientists can synthesise a full genome then an even greater degree of manipulation might be possible. And it might be a lot cheaper. The HGP-Write project also has a way to go – practically, we don't yet know exactly how this creating-genomes-and-sticking-them-in-cells process would work. And we have yet to contemplate the scientific and ethical implications. Regarding CRISPR, Prof. Jennifer Doudna and other leading biologists have called for a worldwide moratorium on the use of her genome-editing technology until its ethical implications can be further explored by scientists and the public. So for today the level of our knowledge of how genes control animal form and function isn't quite there scientifically or ethically. Ah, well. Seems we'll have to stick to burning unlucky witches like Mirri Maz Duur on beloved spouses' funeral pyres to get our dragon eggs to hatch for a bit longer yet. *Sigh* . . .

∽ Mermaids ∽

The world of *A Song of Ice and Fire* is choc-o-bloc with weird and wonderful creatures, and whilst the majority of these are land-based, some make their home in the sea. Take, for instance, that half-fish, half-human classic, the mermaid. Mermaids or 'merlings' crop up in various guises throughout *Game of Thrones*. There's a popular fan theory that Spymaster Varys is a merling – it would explain how he gets around the Seven Kingdoms so quickly. Most notably, The Grey King, ancestor of Theon and the Greyjoys, makes the curious decision to marry a mermaid (thus, we imagine, leading to considerable confusion on his wedding night). However, mermaids also lend a helping hand to the Ironborn who drown in the sea, and Elenei (the wife of Ironborn Durran Godsgrief) appears to have been a mermaid before becoming a mortal, which is said to account for some of her odd behaviour. (Durran would gallantly hold his hand up and proclaim 'Not for Elenei' whenever a trip to get fish and chips was suggested).

But do mermaids exist in the real world?

Although many dismiss the concept out of hand, there are several reasons to remain open minded about the issue.

Almost three-quarters of the Earth's surface is covered with water and, according to the 2010 Census of Marine Life, these vast oceans contain almost two million different species of marine life. Many of these animals are as strange as merfolk (the correct term for mermaids and mermen). Take, for example, the seahorse. Evolution has dealt these curious creatures a deeply odd hand, including eyes that move independently, strange snouts that allow them to creep up on prey undetected, and prehensile monkey-tails that are super-strong and can be compressed by 60% without suffering any permanent damage. Given such genuine oddities, it seems quite possible that a half-human, half-fish might actually exist in one form or another.

Mermaids are not a modern invention but can instead be traced back many thousands of years. For example, in Babylonian times, several sculptures and seals (the sort that you put on documents, not the ones that balance balls on their noses) depict the deity Era, a fish-based god with the upper body of a man and the lower body of a giant fish. Similarly, in Greek mythology the god Triton was a merman who allegedly emerged from the sea bearing important tidings (including the classic 'Jeez, these trousers are killing me').

The appearance of such creatures was often far from good news, with many ancient texts reporting that mermaids were a symbol of ill fortune. However, such negative connotations are not universal. In British folklore, for instance, mermaids are often seen within a more positive context, with some stories describing how they befriended and sometimes even married humans (thus possibly explaining the existence of weatherman Michael Fish and politician Alex Salmond).

Of course, sceptics tend to read about these ancient legends and dismiss them as fantasy. This might, however, be throwing the mermaid out with the bathwater because over the years there have been several sightings of supposedly genuine mermaids. For instance, in one story from the 1600s, eyewitnesses claimed that a mermaid emerged in a Dutch dike, integrated into the community, and even converted to Catholicism. In another report, a 17th century sea captain was sailing off the coast of Newfoundland when he came across a mermaid. The captain later described his experience, noting that the creature had large eyes, a short but finely shaped nose, long but well-formed ears and shocking green hair. He apparently started to fall in love with his oceanic catch before realising that her lower half was all fish.

Perhaps encouraged by such seemingly fantastical tales, hoaxers in the 1800s produced several fake skeletons of mermaids. In the 1840s one of these hit the headlines when the legendary showman PT Barnum announced that he would be exhibiting the 'Feejee Mermaid'. To many, this grotesque figure was half-monkey and half-fish – an obvious fake. But the exhibition was a huge success and thousands flocked to see it. In the end, Barnum's museum was jammed with people, forcing Barnum to hang up signs announcing 'This Way to the Egress.'

Not realising that 'Egress' was simply another word for 'Exit', people followed the signs in the hope of seeing this extraordinary new attraction, only to find themselves out on the street.

Strangely, eyewitness reports of mermaids are not confined to the long lost past. In 2009, for example, several people claimed to have seen a mermaid just off the coast of Israel. According to newspaper reports, the mermaid carried out a couple of tricks and then vanished into the sea. The local tourist board offered a million dollars for the first photograph of the creature, but the mermaid proved remarkably (perhaps predictably) camera shy.

It's entirely possible that such strange sightings have their basis in several sea-based animals, including manatees and dugongs (also known as sea cows). These strange marine mammals are now endangered species. They live in warm, shallow, coastal waters, estuaries and rivers, and possess flat tails and flippers that look a little like (slightly stubby) human arms. The match is far from perfect, but add in the effects of distance, poor weather conditions, low light and several shots of rum, and suddenly they become possible frontrunners for many a sighting.

So there we go. Many of the mermaid-based *Game of Thrones* references are inspired by stories from folklore, genuine, eyewitness reports and science. In the same way that mermaids help the Ironborn after their watery deaths, so many cultures view mermaids as a symbol of good luck. So next time you come across one, don't dismiss them as nothing more than a fisherman's tail, because there's more to mermaids than meets the eye.

CHAPTER 9

HOW WILL IT END?

Where we put on a sweater and consider
astronomical answers to the perplexing question
of a long winter, and attempt to finally settle
the question posed by poet Robert Frost.

∽ Will It All End in Fire or Ice? ∾

Fire and Ice by Robert Frost

Some say the world will end in fire,
Some say in ice.

From what I've tasted of desire
I hold with those who favor fire.
But if it had to perish twice,
I think I know enough of hate
To say that for destruction ice
Is also great
And would suffice.

George RR Martin has said that the great, overarching title of his Game of Thrones novel series, *A Song of Ice and Fire*, was in part inspired by this poem. A lesser-known story is the poem's connection to the American astronomer Harlow Shapley, who is celebrated for working out the size of the Milky Way, and also theorising about where icy planets could be found in the universe. Shapley was the doctoral supervisor of Cecilia Payne-Gaposchkin, who in turn worked out what the sun is made of via the spectra from its flames, and how to measure the elements that make up other stars from how they shine.

In 1960 Shapley was asked by the poet Robert Frost how the world would end. Shapley was *the* astronomer of his day, Frost *the* poet (he won a Pulitzer prize for the collection 'Fire and Ice' appeared in), so this was quite the conversation.

The resulting poem is a tricky work to pin down. It's a very simple verse, until you look closer. The world will end – but which 'world'?

Reading through the lines, you get a vertiginous feeling; we're simultaneously wondering about the fate of our whole planet and a single human heart. '*I* will not die but the world will end,' said Frost's contemporary Ayn Rand the Russian-born American novelist and philosopher (with typical self-effacement.) Do we know what will happen to the Earth in the end? Do we know what will happen to us?

This dizzying perspective seems to echo through *A Song of Ice and Fire* – we feel our way through a vast and complex world, but each chapter of the novel is seen through the eyes of a single character.

George RR Martin has explicity said that the seasons of Ice and Fire are controlled by magic. But such elemental forces also shape the climate of our real world. Could they offer us clues about the battles to come? Is there any astronomical explanation behind the summers and winters in Westeros going on so long? – they are unpredictable and can last for several years. Presumably, as on Earth, a year is defined as one orbit of the planet around its sun or – if the inhabitants still believe that the sun goes around the Earth – by one complete circuit of the sun through the background constellations of the zodiac.

What Causes the Seasons?

I spoke to Dr Marek Kukula, Public Astronomer at Greenwich Royal Observatory, to discover more about the astronomical phenomena behind our seasons.

Earth's seasons are the combined effect of our planet's axial tilt and its annual orbit around the sun. Both the tilt and the orbital path are relatively stable, ensuring that our seasons are predictable and regular. However, if either of them were to vary then the pattern of seasons could become both more extreme and more chaotic.

To a large extent, the presence of our moon helps to stabilise the Earth's rotational axis, keeping its angle of tilt constant to within a couple of degrees and conserving the planet's seasonal patterns of warming and cooling. These will still vary over millions of years due to continental drift and consequent shifts in the distribution of landmasses and oceans.

Earth's orbit is also very close to circular, ensuring that the total amount of heat and light received by the planet remains fairly constant throughout the year. This is not the case for some other planets, such as Mars, Mercury and many exoplanets, which have more elliptical orbits.

～ Solar Variability ～

Lets tour the solar system with Dr Kukula to discover more . . .

The Earth's climate is powered by the energy it receives from the sun in the form of heat and light, so any changes in the energy output of the sun will also affect our climate. The sun is actually quite a stable star but, even so, the turbulent processes inside it do result in a certain amount of variability. Most obviously, activity on the solar surface – including sunspots, solar flares and explosions of gas known as coronal mass ejections – increases and decreases over an 11-year cycle, and this does have a measurable effect on the Earth's climate. However, the effect is very small – much smaller than seasonal variations in temperature – and, unlike the seasons in *Game of Thrones*, its 11-year period is relatively reliable and obvious.

In the late 1800s, by studying records of sunspot numbers spanning several centuries, Greenwich Royal Observatory astronomers Walter and Annie Maunder noticed that there was a period between 1645 and 1715 when sunspot activity was virtually non-existent, even at the expected peaks of the 11-year cycle. Interestingly, this period of suppressed activity, now known as the 'Maunder Minimum', seems to coincide with the most intense phase of the Little Ice Age, a period of abnormally cold winters in Northern Europe.

It is still unclear whether there is a direct connection between the lack of activity on the sun and the change in continental climate, but one theory proposes that enhanced solar ultraviolet output during a period of suppressed sunspots would be preferentially absorbed in the Earth's upper atmosphere, causing it to expand, and in turn diverting the high-altitude jet stream which shepherds weather systems across the North Atlantic to Europe. If a similarly localised process occurs in the world of *Game of Thrones* this might explain why the colder conditions predominantly affect the continent of Westeros.

The internal physics of the sun are still not fully understood and so the processes that triggered the Maunder Minimum remain unpredictable. Other similar events – the Spörer Minimum (1450–1540) and the Dalton Minimum (1790–1820) – have also been identified, but with no obvious pattern. The relative weakness of the most recent 11-year sunspot cycle has even led some solar physicists to speculate that we might be about to enter another prolonged solar minimum, although this idea is controversial.

The most extreme example of solar activity in historical times actually sounds like rather a refined *soirée*. During the Carrington Event of 1859, a tremendous solar flare on the sun launched a huge coronal mass ejection directly towards Earth. When the cloud of high-speed plasma struck the Earth's magnetic field it triggered powerful auroral displays in the upper atmosphere; the northern and southern lights were visible even in places close to the equator. As the shockwaves passed through our planet's magnetic field, electric currents were generated in the long-distance cables of the telegraph network, stunning operators with electric shocks. If such a flare occurred today says Dr Kukula, our satellite, phone and power networks would be incapacitated.

Like the seasons in *Game of Thrones*, solar activity is unpredictable. So we can imagine *their* sun as more variable and temperamental than our own, leading to shorter, more intense and more frequent sunspot minima that correlate with the erratic seasons of Westeros. Some stars can display even more violent activity – 'superflares' up to a million times as powerful as the Carrington Event.

The world of *Game of Thrones* is a pre-industrial society where technology seems to have remained at a medieval level for thousands of years, and perhaps it's just as well. If the *Game of Thrones* sun is indeed a highly active star then erratic climate variations wouldn't be the only problem. Any civilisation reliant on electric power and satellite communications would suffer constant catastrophic setbacks due to solar outbursts. But luckily ravens aren't affected by coronal mass ejections. They may give an extra 'caww' but they keep on flying.

∽ Comet Con ∾

It's widely accepted that the collision of a comet or asteroid with the Earth 66 million years ago was the main factor in the extinction of the dinosaurs. Vast clouds of dust generated by the impact would have spread out across the upper atmosphere, enveloping the globe and drastically reducing the amount of solar radiation reaching the ground. This would have plunged the whole Earth into a cold, dark 'impact winter' lasting for months or even years.

Dr Kukula tells me that some scientists have suggested that less devastating collisions might be responsible for various, localised climate changes – such as the Younger Dryas, a marked cooling of the northern hemisphere around 13,000 years ago – but this is still controversial. And given the utter devastation that a comet crash causes, people would presumably have noticed such an occurrence in *Game of Thrones* if every Westeros winter was triggered by such an impact.

Still, there might be a gentler way in which comets could plunge a planet into a prolonged winter. Each time it swings close to the sun on its elongated orbit, the icy surface of a comet is warmed by solar radiation, causing it to eject vast clouds of dust and vapour that go on to form a tail millions of kilometres long. Long after the comet itself has returned to the outer regions of the solar system this dust hangs around, close to the sun, and whenever a planet passes through it the larger particles burn up in the atmosphere, producing a meteor shower.

If a particularly dense clump of dust is encountered, the resulting shower can be spectacular, like the Leonid shower of November 1833, which lit up the sky like a firework display for several hours. However, the smaller grains can enter the atmosphere intact – and in sufficient numbers – they could conceivably go on to have a cooling effect on the climate.

In 1705 the English mathematician and astronomer Edmond Halley calculated the orbit of his famous comet, working out that it takes 76 years to complete one orbit around the sun. But before that comets were considered to be mysterious and unpredictable visitors.

In *Game of Thrones*, the approaching winter is heralded by the appearance of a distinctive red comet in the sky. This comet, or 'bleeding star', also heralds the return of magic and dragons to the world of course. In astronomy, reddening is often the result of light passing through large amounts of dust. Perhaps the World of Ice and Fire exists in a solar system with numerous comets rich in very fine dust. Astronomers have already detected several systems in which comets are many times more common than they are in our own. Any planets in these systems might be subject to heavy bombardment as well as unpredictable variations in climate due to lingering comet dust.

~ Everyone's a Winner ~

As the sun and the other stars orbit around the centre of the galaxy, they periodically pass through the denser regions of stars, gas and dust that make up its spiral arms. When this happens there is always the possibility that they might pass through a 'giant molecular cloud' – hundreds of light years across – thick with gas and dark space dust.

Normally, the constant outward pressure of the sun's solar wind keeps this stuff out of our solar system, but inside a giant molecular cloud some of the gas and dust would inevitably penetrate, coming between the Earth and the sun and blocking some of the solar heat and light. The resulting decrease in solar radiation reaching the Earth would have a cooling effect on our climate and might cause it to flip into a more wintery state according to Dr Kukula.

Passage through the cloud might take millions of years and the amount of dust-induced screening could vary year by year as the solar system passed through regions of increasing and decreasing dust density – an unpredictable journey that would leave its mark in the form of constant changes of climate on the system's planets.

Ultimately, when our world ends it will probably go the same way as Mercury and Venus – swallowed by our expanding fiery sun. But it's possible something of our Earth could survive, even millions of years into the future when our sun has become a red dwarf. Water molecules left hidden in the rocks that once made up Earth will break up and float through the universe. One day they may coalesce and grow, and over millions of years new planets will form, warmed by a new sun, but cooled by the water that was once ours. Who knows what will happen to this new planet, what stories will be told there? Ice and Fire could both win.

ACKNOWLEDGEMENTS

Firstly of course I'd like to add my thanks as a fan (alongside millions of others!) to George RR Martin for creating such an utterly wonderful, rich, engrossing world.

To my marvellous agent Susan Smith – thank you for believing in me! To the wonderful teams at Coronet, Hodderscape (or Hodorscape, as they will always be to me) and Little, Brown.

It was a huge privilege to get so much help and learn about fascinating science from all the experts I talked to for the book – a few were friends, some acquaintances, but many were complete strangers and all gave their time, insights and knowledge to this project with great generosity. Please do go to www.helenkeen.com/morescienceofGoT to find out more about them.

I'd particularly like to thank the following founts of wisdom: Ryan Consell – for all the swords and grand armour I now keep in the Great Hall of my imagination, Dr Marek Kukula, Dr Jonathan Pettitt, Dr

Kelly Weinersmith and Professor Richard Wiseman for much magic and kind encouragement.

Thank you to the following friends for encouragement, ideas, draft help, and general brilliance – Ian Simmons, Tim Hemmings, Miriam Underhill (particularly for being the best possible person to watch Season 1 with) and extra huge thanks to Deborah Sabapathy for so much helpful editing and for being an inexhaustible source of kindness, great ideas and even better jokes.

So, one final time, thank you so much one and all. You have my banners!